Praise for *Wildly Successful Farming*

"We don't have to choose between healthy land and productive land—we can have both. DeVore's careful chronicling of Midwest farmers who practice an agriculture that respects and supports nature will give you hope for the future."

Kristin Ohlson,
author of *The Soil Will Save Us*

"Most Americans have forgotten that the success of agriculture depends on the wild world it so often displaces. These farmers remind us that those two vital elements need not be mutually exclusive—indeed, the success of food production depends on a healthy natural world."

Lisa M. Hamilton,
author of *Deeply Rooted*

"Meet optimistic realists—farmers, conservationists, and scientists—blurring occupational boundaries to reveal a world in which agriculture and ecology are productively intertwined. These are outliers in a sea of corn and soybeans (and resultant depauperate ecosystems), but their stories, told often enough, can change all of that."

Karen Oberhauser,
director, University of Wisconsin–Madison Arboretum

Wildly
SUCCESSFUL
FARMING

Sustainability and
the New Agricultural Land Ethic

Brian DeVore

The University of Wisconsin Press

The University of Wisconsin Press
728 State Street, Suite 443
Madison, Wisconsin 53706
uwpress.wisc.edu

Gray's Inn House, 127 Clerkenwell Road
London EC1R 5DB, United Kingdom
eurospanbookstore.com

Printed in the United States of America

This book may be available in a digital edition.

Library of Congress Cataloging-in-Publication Data

Names: DeVore, Brian, author.
Title: Wildly successful farming: sustainability and the new
agricultural land ethic / Brian DeVore.
Description: Madison, Wisconsin: The University of Wisconsin Press, [2018]
| Includes bibliographical references and index.
Identifiers: LCCN 2018011392 | ISBN 9780299318802 (cloth: alk. paper)
Subjects: LCSH: Sustainable agriculture—Middle West. | Alternative agriculture—
Middle West. | Agriculture—Environmental aspects—Middle West.
| Agricultural landscape management—Middle West.
Classification: LCC S444 .D48 2018 | DDC 333.76/160977—dc23
LC record available at https://lccn.loc.gov/2018011392

ISBN 9780299318840 (pbk.: alk. paper)

For

Earl and **Helen DeVore**

who let me roam after the chores were done

Contents

Illustrations

Acknowledgments

I owe an incalculable debt to all of the farmers over the years who have invited me into their kitchens, barns, and fields and shared that most precious of commodities: time. They patiently put up with my repeated attempts to gather enough insights to fully understand the complex balancing act of producing food profitably while nurturing a working ecosystem. For every farm described in this book, there are at least two dozen more that could have been included given more time and resources. I only hope I've come close to communicating the passion with which these wildly successful farmers go about their daily lives. I've also benefited from the willingness of experts representing various aspects of agriculture, science, economics, and policy to put up with "just one more question." For someone who barely passed a basic soils course in college, access to such expertise is a goldmine.

This book would not be a reality if it wasn't for Dana Jackson. Her poking, prodding, questioning and, most important, support, motivated me to pull together my years of reporting and fashion them into a somewhat coherent narrative. But even more important, her belief in my ability to help tell the story of sustainable agriculture can be traced back to 1994, when she hired me as the editor of the *Land Stewardship Letter*, the official publication of the Land Stewardship Project. That job set me off on a two-decade journey that brought me in touch with some of the most innovative and inspiring farmers I've ever had the pleasure to meet. Finally, she was generous enough to give a draft of this manuscript the kind of tough-love read it required at the time. Thank you, Dana.

I also want to thank Ron Kroese and the late Victor Ray, who in 1982 had the audacious idea of creating a nonprofit organization dedicated to the stewardship of working farmland, as well as the people and communities who depend on it for survival. I am grateful and humbled to have had the opportunity to utilize my skills in some small way as a staff member of the Land Stewardship Project and am constantly in awe at the creativity and tenacity of this organization's dedicated staff, board, and membership. To paraphrase Wendell Berry, they pursue their work with joy, even though they've considered all the facts.

This book's title and theme are the result of conversations and many pasture/birding walks I've had with Art "Tex" Hawkins, whose ideas about blending the wild and the tame have reinforced what I've witnessed on farms across the Midwest. His dedication as a public servant who truly cares about the land and the people traces its roots all the way back to Aldo Leopold himself.

And thanks to Gwen Walker, my editor at University of Wisconsin Press, who immediately "got" what this book's title represented. Her enthusiastic support of the original proposal, along with the patient guidance she, Adam Mehring, and the rest of the University of Wisconsin Press provided made it possible to bring these stories to a larger audience. The anonymous reviewers who gave the draft manuscript a thorough and thoughtful reading improved the final version immensely in myriad ways.

My late parents, Earl and Helen DeVore, passed on to me a love of the land and the community that surrounds it, as well as a respect for the hard work that goes into trying to make a living in agriculture. My older siblings—Rod, Sheila, Sandy, and Sherry—left behind their books when they departed the farm to start their adult lives; these were priceless hand-me-downs for an awkward kid. Speaking of family, a special thanks to cousins Bobby Dale and Charlotte Groves, who shared food, beer, and good company with me while I was holed up writing for a few weeks on the DeVore farmstead. Thanks for reminding me why farmers do what they do.

Ben and Molly, I apologize for all the nights and weekends that took me away from you while I was traveling from farm to farm, trying to track down yet another story. Returning home to your laughter and curiosity after those reporting trips made it all worthwhile. As you begin the next stage of your lives, I hope you can find your own way to blend the wild and the tame.

Finally, none of this would mean anything without Kathleen. Her fierce support, combined with patience and belief in me, held this project together, even when it looked like it was fated to remain a stack of un-organized field notes. Sometimes we farm boys need a little Boston pushiness to get 'er done. From the Maloti Mountains to Minnesota, your dedication to racial justice and education has inspired me to pursue my own passions. Kea u rata.

Wildly Successful Farming

Introduction

A Day on the Farm, a Night on the River

Dan Specht finished up hog and cattle chores, hopped into his pickup truck, and wound his way down to the Mississippi River, just a few minutes' drive away from his hilltop farm in northeastern Iowa. He had fishing gear in the back, soil under his fingernails, and nutrient runoff on the mind. That wasn't unusual for Specht. It was difficult for the farmer to separate his various passions—even if they seemed to come into conflict at times.

"I'm trying to be more efficient in my nutrient cycling," the soft-spoken bear of a man told me that summer evening as he guided the pickup past corn, soybeans, alfalfa, and pastures before hitting the heavily timbered river bottom, which was home to, among other things, the ancient, humped structures of the effigy mound builders culture. "The thing is that corn and beans don't create a very complex rotation. It's a very leaky system. It's annual, warm season row crops, and it's the middle of June before the roots start picking much up. Before you know it, your drain tile lines are running full of nutrients the whole months of April, May, and June."

There, in one succinct description given during a five-hundred-foot drop in elevation that carried us from the cultivated farmlands to the wild bottoms of the Upper Mississippi, Specht had laid out to me perfectly a problem that touched on plant physiology, soil biology, and hydrogeology. And because his beloved Mississippi flows down to the Gulf of Mexico, where all that nitrogen that's escaping leaky farm fields has

3

helped create an oxygen-poor "dead zone" that, as of 2017, was the size of New Jersey (approximately nine thousand square miles),[1] Specht's description demonstrated he also understood Barry Commoner's first law of ecology: "Everything is connected to everything else."[2]

This conversation took place in 1999. I had called Dan and asked if I could visit him for a day to talk about the Gulf of Mexico dead zone and the role agriculture plays in it. The dead zone, which has decimated fisheries in the Gulf, has its roots in a midwestern farming system that has increasingly become dependent on monocrops of corn and soybeans. Raising corn, for example, requires heavy dosages of nitrogen fertilizer, and much of it—20 percent or more in some cases—escapes down into the soil profile, making its way into field drainage systems and eventually down the Mississippi to the Gulf, where it supercharges algal growths that gobble up oxygen.[3] I've looked at historical charts that illustrate a clearcut mathematic equation: more corn + more fertilizer = less life in the Gulf of Mexico.[4]

I had read all the government studies on the topic, and perused the statements issued by the environmental community ("Farmers are killing the Gulf!") as well as the agribusiness industry ("We can't feed the world without nitrogen fertilizer!"). But to write an article that went beyond the science and the rhetoric, I needed someone to give me a ground-level view of the situation. Dan represented a way to put a human face on an incredibly complex and controversial topic.

I had met Dan a few years earlier at a sustainable agriculture workshop and was impressed with his knowledge of not just farming but the landscape it was set in. He was an avid hunter and angler, had studied wildlife biology at Iowa State University before leaving to go into farming fulltime, and later in life got a biology degree from the University of Northern Iowa. He was a bit of an expert (in a good way) on everything from local springs and karst geology to the birds that called his corner of Iowa home. It was all part of a blend, and Dan was the kind of guy who surprised you: after pulling a few words of small talk out of him, suddenly you realized you'd just gotten the entire history of the Big Spring Project, a pioneering research initiative in his community that set the standard for learning about and mitigating agricultural fertilizer pollution.[5]

And he walked the talk: Dan wouldn't dream of doing anything to upset the delicate balance between farming and the land that he felt should exist. I'll admit, at times Dan took blurring the lines between humanity and nature a little too far. Once, while I was eating a bowl of ice cream in his somewhat rustic kitchen, he opened his back door and grabbed a handful of mulberries that were sprouting from a limb scratching at the side of the house. He threw them into my bowl, stems, ants, and all. The resulting concoction was delicious, if somewhat more fibrous than I had bargained for.

Not long before I had visited Dan's farm for the first time in 1999, he traveled to the Gulf of Mexico as the guest of an environmental group, getting a firsthand look at the impact of agricultural pollution on the people and their livelihoods. The experience had reinforced his commitment to reducing the amount of nutrients leaving the five hundred acres he farmed at the time. Many farmers would approach such a problem in a reductionist manner: too much fertilizer escapes my land, so I will put in place a specific practice or structure to control it. Such thinking has resulted in innumerable terraces, controlled drainage systems, and even "bioreactors" (utilizing material such as wood chips to soak up excess nutrient runoff) on farms across the Midwest, often at taxpayer expense.[6]

They've had mixed results. Such practices have helped reduce pollutants on a local basis, but on a watershed-wide level, we still have major problems, and not just in the Gulf of Mexico. In 2015, the Des Moines Waterworks sued three northwestern Iowa counties, claiming drainage districts there act as conduits for nitrate to move from farm fields into the Raccoon River, a major source of water for five hundred thousand residents. Such contamination has forced the city to invest massive amounts of money in equipment just to make the water safe for drinking.[7] Agricultural runoff led to massive algal blooms in Lake Erie during 2014. As a result, for three days Toledo, Ohio, had to shut down the drinking-water system that services four hundred thousand people.[8] Such problems are caused by nonpoint source pollution runoff, which is particularly difficult to control since it comes from numerous places on the landscape, rather than one specific "point" source such as a storm sewer pipe emptying straight into a river. The latest U.S. Environmental Protection Agency National Water Quality Assessment shows that agricultural nonpoint source pollution is the leading source of water quality problems on surveyed rivers and streams, the third largest source for lakes, the second largest source of impairments to wetlands, and a major contributor to contamination of estuaries and groundwater.[9] Climate change, which is bringing torrential rains to parts of the Corn Belt, is exacerbating the problem. In 2017, the journal *Science* published a paper showing that increased rainfall could increase nitrogen runoff as much as 20 percent by the end of this century, which would slash oxygen levels even more in not just the Gulf but places like Chesapeake Bay.[10]

As farmers like Dan Specht see it, the struggle to contain nonpoint source pollution is an indicator that stopgap conservation structures don't encompass the big picture "system" approach to farming in concert with the land. Many of these mitigation measures are steeped in a basic mindset: How can we continue to raise corn in a manner that will keep the regulators off our backs? That's the kind of thinking that predominates when people see themselves as "corn producers" only, rather than farmers who are willing, depending on the circumstances, to consider raising a variety of products.

Dan had the ability to raise a high-yielding crop of corn. But to him, such a crop was a means to an end, not an end itself. Therefore, he approached the leaky nutrient problem much more holistically. He asked such questions as, "Should I be raising corn on this particular piece of ground in the first place? Rather than raising corn and selling it to the local elevator so that it could eventually be fed to livestock, why not raise livestock on that land myself?" This kind of thinking led Dan to do such things as produce beef cattle on his hilliest acres utilizing a system called managed rotational grazing. Developed in France and New Zealand and modified to fit local situations, this system consists of moving livestock through a series of grazing paddocks on a regular basis—sometimes as much as once or twice a day—so that they don't overgraze the pastures. It distributes manure and urine across the landscape evenly, providing grasses and forbs an opportunity to take up the nutrients at a sustainable rate that fits their needs. Because it eliminates overgrazing, such a system can extend the pasture season by a month or more in the Upper Midwest, which is a financial bonus for farmers, particularly beef and dairy producers.[11]

As managed rotational grazing has caught on in this country, farmers and scientists have noted numerous other benefits: it sequesters greenhouse gases, while providing habitat for grassland songbirds and pollinators.[12] One other important benefit is that such a system can be set up at a lower cost than, say, a full confinement livestock facility reliant on high inputs of machinery, energy, and drugs. As a result, managed rotational grazing has provided an entrée into livestock production for many cash-strapped beginning farmers in recent years.[13] Since managed rotational grazing provides an economically viable reason for keeping the land covered in perennial plants such as grass, it can be a way to counter the trend of more and more acres going under the plow to grow annual row crops like corn and soybeans.

Jeff Klinge and the late Dan Specht fishing on the Mississippi River near their farms in northeastern Iowa.

I spent the day on Specht's crop and livestock farm and saw first-hand how he utilized rotational grazing on his steepest fields while bobolinks and bluebirds flitted about in pastures surrounded by oaks. Most of Iowa is former tallgrass prairie and in many parts of the state any field that dares to rise even a few feet above the surrounding landscape is considered mountainous. But Dan's neighborhood is part of the "Driftless Area"—a region dominated by rugged bluffs that were not shaved down by the last ice age. Some of his fields are so steep that people joke that squirrel hunting on this land involves aiming a .22 rifle *down* at the tree canopy. Dan reserved his flatter acres for raising corn and soybeans, and even in his row-cropped fields he utilized diverse rotations and soil-building cover cropping to keep nutrients on his land and out of the water. Such methods also kept topsoil in place. That's not an easy task; during my first visit I noticed how a recent rain had washed soil and plant debris off a neighbor's field, forming a dirty, stucco-like wall that plastered a fence line separating it from Dan's property.

The fishing trip we took at the end of that informative day was not just a way to blow off steam over a few beers—to Specht it was part and parcel of the personal seminar he was giving on farming, fishing, and fertilizer. It became clear as he, neighboring farmer Jeff Klinge, and I cast lines and talked about everything from agricultural policy to water chemistry to geology, that there was no divide between what took place up on the nearby hilltops and the results down on the bottomlands, all the way downstream to the Gulf, almost two thousand miles away. It was all interconnected.

"It's really fragile," Dan said at one point while a freight train rumbled along the Wisconsin side of the river. "It's vast, but it's fragile." He was referring to the point where the Mississippi meets the Gulf, but it was clear he had his own neck of the woods in mind as well.

Making Connections

I thought about that jam-packed day more than a dozen years later while standing on a sidewalk in St. Paul, Minnesota. I'd just gotten the news that Dan had been killed at the age of sixty-three in a haymaking accident on one of those steep fields that made up his farm. I had spent the intervening years writing about other farmers that refused to separate food production from ecological processes, and Dan had been the spark that ignited my interest in this kind of agriculture.

Frankly, farmers like Dan Specht are not the norm. I've seen innumerable examples of farms that place a wildlife pond here or a windbreak there in the name of "conservation." Every year farmers throughout the Corn Belt are given awards by commodity groups or Soil and Water Conservation Districts for owning and operating some version of an "Outstanding Conservation Farm." Invariably, when the press release is issued it provides a shopping list of practices and structures the winners have put in place: a wildlife planting or grassed waterway, planting cover crops to protect soil between the regular cash crop growing seasons, switching to no-till to avoid exposing fields to erosion.

These practices are all well and good, and when taken together make an impressive scorecard. There is no doubt, for example, that no-till farming, which does away with using the moldboard plow to disturb the land, has saved billions of tons of soil in recent decades.[14] But these disparate practices don't always add up to an integrated system, and thus are vulnerable to being dropped as soon as market, regulatory, or other incentives disappear.

But occasionally over the years, I'd run into someone like Dan Specht, who wasn't just cherry-picking a practice here and there in order to provide a short-term solution to a problem. They have a comprehensive view of a world where agriculture and ecology are deeply interconnected, and realize any real, long-term sustainability must be rooted in taking advantage of these connections in a positive way. Such a

Dan Specht did not differentiate between the "wild" and the "domesticated" parts of his farm.

worldview is not easy to come by. I know, because it didn't come natu-rally to me.

Growing up on a 240-acre farm in southwestern Iowa during the 1970s, I considered farming and the natural world to be two very different animals. You raised corn, soybeans, hogs, and cattle up on the DeVore hill, and wildlife thrived in those hidden, and somewhat mysterious, hollows down in the bottomland where a scrappy little stream called the 7-Mile cut a gash through our neighborhood. That belief was reinforced by the fact that the 1970s was witness to a "fencerow-to-fencerow" grain production explosion, when farmers were encouraged to farm every last acre in the name of "feeding the world."[15] Wildlife habitat was a luxury on "real" working farms. My home place succumbed to this thinking, but not nearly to the extent of other farms. My dad removed plenty of trees and a few brushy fencerows—more out of a need to see things neat and tidy than any desire to "feed the world." He had hunted, fished, and trapped back in the day, but now he was a farmer, and his interaction with the land began and ended with how to wrest a living from it.

I had always preferred spending more time in the untamed bottoms of the 7-Mile than the domesticated tops of our farm, so when I went to college I studied fish and wildlife biology and journalism, thinking I was going to be the next Mark Trail, the outdoor writer of newspaper comic strip fame. My belief that farming and the natural world did not mix was reinforced early by a professor who was a big believer in the role wildlife refuges, parks, and other publicly owned natural areas could play in preserving remnants of nature. During a field course called "The Ecology of the Missouri Ozarks," we drove by section after section of farmland so we could visit federal and state wildlife refuges, waterways that had been protected via the Wild and Scenic Rivers Act, and national forests.

At one point, we visited a Missouri lake that was inundated with boaters and swimmers on an early spring day. It was a cacophony of activity, and quite frankly, did not show American outdoor recreationists at their best. It was particularly jarring after spending several days in the more "natural" parts of the Ozarks.

"We need areas like this," my professor explained as he slowly guided our college van through the crowds. "It's like a safety valve that keeps people away from real nature." By extension, farming the hell out of the best soils kept farmers away from those "natural areas" as well, I assumed. This would mark a common belief I ran into in subsequent years: it wasn't just farmers who thought agriculture and the natural world did not mix, so did many ecologists and government natural resource professionals.

And during my college years in the early 1980s, farming and nature seemed more alienated from each other than ever. I wrote articles about studies showing that agrichemical contamination of Iowa's drinking water wells was ubiquitous, while wildlife habitat was shrinking to all-time lows. I'm reminded of the first magazine story I tried to write as a student journalist. It was supposed to be about how farming could actually help wildlife. My first interview was with a well-known bird expert at Iowa State University. When I presented him with my thesis for the story, he said something along the lines of, "I'm not sure if there's a story there. Farming does more harm than good to wildlife." He then went on to talk about how harvesting corn with combines left the fields so denuded of grain that there wasn't even anything left over for the wildlife to glean. I ended up doing the story anyway, but it was pretty lame—something about farmers planting wildlife-friendly windbreaks along fence lines sticks out in my head. To be sure, this was 1983 or so, and we had just come through the 1970s and its decimating fencerow-to-fencerow planting fever. That ornithologist was right to be pessimistic.

It turns out there was a very important human element to all this bad news that I missed at the time: as farms became less numerous and larger, environmental degradation increased. There were simply fewer people on the land to care if a pasture was plowed or a brushy fence line bull-dozed. It turns out the fate of the family farmer isn't just tied to the price of corn—there is a real connection to the health of the land as well.

Somewhere along the line, my view of agriculture's relationship with nature changed. I'm sure it was a combination of things, but one experience stands out. One of the farmers who farmed on the bottomlands of the 7-Mile had hunted, fished, and trapped with my dad back when they were young. But this guy, perhaps because of his nearness to the creek, had never quite lost his interest in woods and streams. One day he told me about a book he had just read. "You might be interested in it," he said nonchalantly. "It's about this guy who kind of takes you through the seasons and describes our relationship with the land, things like that."

That book was *A Sand County Almanac*, by Aldo Leopold. Even though it had been written almost half a century before I picked it up, it set my brain on fire. I loved the descriptions of nature, and Leopold's ability to observe one small aspect of the land and extrapolate it into a larger way of thinking. But what really drew me in was his "land ethic"—the idea that we have a moral responsibility to the land and its wild flora and fauna, that not every "cog and wheel" in nature must justify itself economically in order for us to give it permission to exist.

"A thing is right when it tends to preserve the integrity, stability, and beauty of the biotic community," wrote Leopold. "It is wrong when it tends otherwise."[16]

When I first read those words, I didn't know enough about "biotas" or "ecosystems" to judge what land practices were preserving "integrity and stability." All I knew was I didn't like the cold feeling I got in my gut when I saw a stand of hardwoods bulldozed and burned, a creek straightened, or a pasture plowed up and planted to corn. To be honest, when I was younger my opposition to such "land improvements" was rooted in self-interest. I remember when a neighboring farmer ripped out a brushy fencerow soon after he bought property in our neighborhood. My first thought was, "Well, I'll have to find another place to hunt rabbits."

But Leopold helped me begin observing the workings of the land less as a consumer of outdoor diversions and more as a member of a community, one that was much more interesting than I could have imagined. When we stop viewing land as mere property, the possibilities, for us as well as that land, are opened wide. Ironically, some of those possibilities are actually based in economics. For example, farmers in recent years are discovering that by ignoring all that "useless" soil biota and focusing exclusively on adding to their fields' financial value with high-priced, artificial inputs like petroleum-based fertilizer, they are sacrificing the long-term viability of their land.[17] That would not have surprised Leopold, who called it a false assumption that "the economic parts of the biotic clock will function without the uneconomic parts."[18]

And I liked what Leopold said about how an all-encompassing ethic should apply not only to pristine wilderness areas but to where we live and work every day. When one considers such an ethic in terms of agriculture, the health of the land in rural areas is best served when food production and wild areas exist side-by-side, rather than as separate entities performing seemingly unrelated tasks. Such a way of looking at the world was highly appealing to a wildlife-loving farm kid who was living nowhere near a national park, wildlife refuge, or nature preserve. If I was to have any interaction with nature, it had to be in the pastures, crop fields, and ditches that made up my agrarian world. Seeing my family's 240 acres through Leopold's eyes suddenly made that farm seem much larger and layered.

In his essay "The Farmer as a Conservationist," Leopold eloquently described how woodlands, meadows, sloughs, and wetlands, those odd corners where ecological services quietly go about their business, can coexist with corn production, pasturing, and other farming enterprises.[19] Wilderness areas, national forests, and wildlife refuges are important. But as Dana and Laura Jackson point out in a book I contributed to in 2002, *The Farm as Natural Habitat: Reconnecting Food Systems with Ecosystems,* too often people see their presence as an excuse to sacrifice a functioning ecosystem on good farmland: "Farm the best and preserve the rest."[20]

The result of this segregation on a landscape scale is pristine preserves such as the Boundary Waters Canoe Area Wilderness on one end of the spectrum and ecological sacrifice zones such as the Corn Belt on the other end. On an individual farm scale, it often means gradual elimination of residual habitat fragments on the assumption that displaced wildlife can simply take up residence on public land somewhere else. The whole concept of wetland mitigation fits this mindset—in a state like Minnesota, highways, housing developments, and other such projects can often be allowed to displace a natural wetland as long as developers provide the means for a replacement marsh to be established elsewhere.[21] It's a nice idea, but ignores the concept that perhaps there was an important ecological reason that wetland was located in its original spot.

An integration of the tamed and the wild not only makes economic sense by saving soil and protecting water quality, for example, but it provides a certain "wholeness" that is so critical to the overall success of a farm. Wrote Leopold: "No one censures a man who loses his leg in an accident, or who was born with only four fingers, but we should look askance at a man who amputated a natural part on the grounds that some other part is more profitable."[22]

In the seven decades since Leopold wrote those words, it has become clear he was right in more ways than one. The sustainable agriculture movement is based on the idea that all aspects of a successful farm—from its soil, croplands, and pastures to its woodlands and sloughs—are part of a healthy whole. Farmers and scientists are realizing that an agricultural operation too far removed from its biological roots is more vulnerable to disease, pests, and uncooperative weather—in other words, it's less resilient.[23]

Leopold was writing in a different era, when industrial agriculture and agroecological thinking were both in their infancy. But recent research and real-farm experience have proven him right. Environmentalists are now aware that creating islands of natural areas is not sustainable in the long term. To be sure, waterfowl benefit from state and federal wildlife refuges, but when migrating they rely on the food and shelter present in the potholes and sloughs found on farms across the Midwest. A protected waterway may be safe from having factory waste dumped straight into it, but what about the non-point runoff from all the farms present in the surrounding watershed? It can be ecological death by a thousand hidden, and not so hidden, cuts.

In places like the Midwest, working lands conservation (of course, it could be argued that all lands "work" in terms of ecosystem services) is

more than a nice concept—it's a necessity in a region where vast tracts
of publicly owned acres are few and far between. Pore over a map where
shades of green indicate the percentage of land devoted to agricultural
production and you'll see the midsection of America resembles the
dominant clothing choice for a St. Patrick's Day parade. In Iowa, almost
89 percent of the land area is farmed. Even in states like Minnesota and
Wisconsin, with their timberlands and lakes, 54 percent and 45 percent
of the landscape, respectively, is in agriculture. Nationally, privately
owned croplands, pastures, and rangeland make up about 40 percent of
the terrestrial surface area, and are managed by just 2 percent of the
population.[24]

And a tiny subset of that 2 percent makes support of a functioning
ecosystem a priority when producing food. But when one gets an opportunity to see such a philosophy in action, the psychological impacts can
have an outsized effect. Soon after I went to work for the Land Stewardship Project in the mid-1990s, I had the opportunity to encounter
numerous farmers, such as Dan Specht, who blended the natural world
and their farming systems almost seamlessly. As I wrote in *The Farm as
Natural Habitat*, some of these farmers were prompted to make major
changes in their operations by health concerns (a well contaminated
with agrichemicals), while others were triggered by economics (seeking
a premium price in the organic market, for example).[25]

I've always been fascinated by what prompts a farmer to make
changes that are likely to invite the derision of others in the community,
particularly other farmers. But the stories I find particularly intriguing
are those of farmers who I call "ecological agrarians"—people who
never really separated the natural world from food production. Sometimes they seemed to be born with this inability to disconnect the two.
Other times, early life experiences forged this connection.

Wildly Successful Farming

No matter what the avenue or the timing, the result is, as Art "Tex"
Hawkins refers to it, "wildly successful farms." Hawkins is a former U.S.
Fish and Wildlife Service watershed biologist who went on to start a
sustainability initiative at Winona State University in Minnesota, and
his late father, Art Sr., was one of Leopold's first graduate students. The
younger Hawkins has worked closely with many of these farms that are
blending nature and agriculture, helping them, for example, monitor

the health of the ecosystem via bird identification. He sees them as the living embodiment of what good can come from refusing to separate the ecological from the agronomic.

I like the term "wildly successful," partly because it's a play on the title of a well-known farm magazine, *Successful Farming*. Just as a pop musician pines to be on the cover of *Rolling Stone*, it's long been acknowledged in the agricultural world that to have your farm featured in a magazine like *Successful Farming* is a sign that you've arrived. Farm magazines like this offer up lots of practical advice, but like their glossy counterparts in the suburbs and cities, they also have an aspirational component to them. "You too can be a successful farmer!" is the message their profiles and photographs convey.

Such success is measured by how many bushels are being produced on how many acres utilizing what kind of technology and inputs. However, there have always been groups of farmers who measure success based on how well their production systems interact with the land's natural functions. When done right, these farms are able to succeed not only ecologically but financially and from a quality of life point of view. These are the wildly successful farms I refer to in the title of this book.

And we're not just talking about farms that are homes to ducks and deer. "Wildness," in this case, extends beneath the surface as well: healthy soil is perhaps the most diverse ecosystem on earth and maintaining its diversity to the point where natural systems can function has repercussions all the way up the food chain, to us. So, this book isn't just about marshes and prairies—it's about farms that give a variety of natural forces the opportunity to interact with human-driven forces in a positive way, literally from the ground up. Some of those interactions happen through a hands-off approach; others are more directed. Either way, thought and conscious decision-making must go into the process for it to be successful. This is, after all, for all intents and purposes the Anthropocene, a geological epoch dominated by the actions of human beings.

This book tells the stories of farmers across the Midwest who are balancing agricultural productivity with a passion for all things wild. These farmers are utilizing a wide range of techniques and strategies, but they are united by a single philosophy, which is to approach working lands conservation as, to quote Leopold, "a positive exercise of skill and insight, not merely a negative exercise of abstinence or caution."[26]

Whenever people read or hear about a farmer who is doing innovative things to balance food production with a healthy, functioning ecosystem, a natural response is, "Nice story, but what does it mean in the bigger picture?" In other words, are these examples fated to be feel-good tales that have no real relevance when it comes to making our overall food and farming system more sustainable? I don't think so. This book also describes how wildly successful farmers can have impacts beyond their field borders—all the way to research test plots and our supper tables. That said, I don't want to mislead readers into thinking that this book is reporting on some sort of regenerative farming revolution that's sweeping the U.S. Corn Belt. These stories are about farmers who are returning resiliency to their particular piece of the landscape. Given the catastrophic environmental threats our entire region faces, it can be difficult to accept the fact that wildly successful farming is an outlier.

But for now, it is—we will need to seek our optimism in the isolated, but powerful, examples profiled here. I've chosen to focus on the Midwest in this book because in many ways its landscape has been thoroughly reshaped by production agriculture to a greater extent than any other region in the country. If wildly successful farming can get a foothold in an area where, in some counties, 95 percent or more of the landscape is blanketed in either corn or soybeans during the growing season, then there's hope for other places dominated by a version of the "Corn, Bean, Feedlot Machine," as Montana rancher Becky Weed calls it. To use an agronomic metaphor, wildly successful farms are seeds and it remains to be seen whether society provides fertile soil—both in the marketplace and the public policy realm—for them to sprout the kind of growth that spreads widely.

This is no shop manual on what that seedbed should look like, but I've included in these pages what I see as some of the common traits shared by farming operations that have in some way become wildly successful. Teamwork, cutting-edge science, curiosity, a willingness to ignore the conventional wisdom adhered to by peers, the ability to foresee (or at least weather) unintended consequences, and, of course, personal passion all play key roles. That's a complex formula, one that makes creating a standard template for being wildly successful pretty much impossible. But when was the last time something truly transformational came out of a neatly organized toolkit?

1

Beyond the Pond

How One Farm Measures Success

It was high summer, and Loretta and Martin Jaus had taken a break from crop work on their west-central Minnesota dairy farm to stand in tall grass, listening and watching for signs of success on their operation. To the right was a straight-line gash of manmade ditch, the kind that's common in this part of the state. Across the ditch was a cornfield sitting on a former lakebed, made possible by the artificial drainage the ditch provided. But in front of the couple were eleven acres of shaggy wildlife habitat: a mixture of prairie and wetland. It was crackling with the sounds of bird life, including the buzzing *zhee, zhee, zhee* of the clay-colored sparrow, a relatively rare bird that is pretty picky about its habitat requirements.

Martin and Loretta were thrilled. They conceded that back in 1993, when they restored this habitat on prime farmland, there were scratched heads and rolled eyes in the neighborhood. They are just a few miles from Renville County, the state's number one corn and soybean producer.

"Does it make sense financially to take that eleven acres out of production?" Loretta asked me rhetorically as she watched birds flit around after insects. "No, but we need it. For us it just made good sense because it's important for us to have diverse numbers and species of animals and plants on our farm. If the place is good for wildlife, then we know it's good for us."

Martin put it more bluntly: "If those sounds weren't there, we would consider ourselves a failure."

This is a wildly successful farm, plain and simple, but not solely because of those eleven acres of avian paradise. Or because of the small amphibian and dove ponds they've established on odd corners of the property, or the one hundred bluebird houses mounted like tiny sentry boxes on fence posts around their grazing paddocks. (Martin estimates they can get one pair of bluebirds for every five acres of pasture. "The best year was thirty-five pairs," he said once with the kind of pride that other farmers might reserve for talking about bushels-of-corn-per-acre yields.)

What makes the Jaus operation, and others like it, special, is its ability to integrate aspects of a healthy ecosystem into the "working" aspects of the farm. This is not the typical attitude in the agricultural or environmental communities. The argument is often made that profitable farming and quality wildlife habitat, for example, don't mix. If we want to leave areas for birds, mammals, and even frogs, goes this argument, the best thing to do is create isolated wildlife refuges where no economic activity takes place. That way, farmers can be free to intensively cultivate every inch of their operations without having to worry about wetlands, shelterbelts, and grassy nesting areas.

Writing in the journal *Frontiers in Ecology and the Environment*, scientists from the Australian National University and Stanford University provide a description of two dramatically contrasting manners of managing the agricultural landscape.[1] In a "land sparing" scenario, acres are farmed intensively—large-scale monocultural operations raising crops such as corn and soybeans are used to produce high yields, for example. In theory, sacrificing these agricultural acres for food, and increasingly fuel, production, makes it possible to set up nature reserves separate from farms on land that normally could not produce high yields of crops or livestock. Usually these reserves are owned or somehow managed by the government or a nonprofit organization, since they do not produce the kind of income private landowners need. Or, in the case of the Conservation Reserve Program or Wetlands Reserve Program, the government pays the landowner not to farm certain "sensitive" acres.

"Wildlife friendly farming," in contrast, is characterized by interconnecting patches of native vegetation scattered throughout the landscape, as well as a high level of spatial heterogeneity—in other words, a diversity of growing plants in a range of small fields while retaining habitat features within the fields, such as buffer strips or scattered trees along streams, wildlife travel lanes, or field borders.

Even groups like The Nature Conservancy, which has long focused on simply buying up vulnerable land and locking out all economic activity, including farming, is starting to see the limits to such a strategy and the benefits to working with farmers who are utilizing nature friendly methods.

"We realize that there isn't enough money out there to buy up all the land. Besides, people make a living from this land," Neal Feeken, who works on prairie recovery and renewable energy in The Nature Conservancy's Minnesota office, once told me while we hiked the pastures of a farm in the western part of the state. "We need to show economic activity can take place on land that's producing environmental benefits."

Diversity = Stability

The Jaus operation is certainly proving economic activity and environmental benefits can cohabitate. During the growing season, their sixty-cow dairy herd gets its nutrition by grazing on a series of small paddocks utilizing managed rotational grazing. Because the farm is certified organic, the Jauses are able to sell their milk for a premium; the level of that premium varies, but can be as much as double the conventional price. It's a nice financial reward for taking the extra trouble to avoid toxic chemicals and petroleum-based fertilizers, among other environmentally friendly practices, and is allowing the family to make a comfortable living without milking several hundred cows and cropping thousands of acres. The fact that neither Martin nor Loretta are working off the farm is significant—it's rare these days to find a family farm where at least one spouse doesn't have a "town" job to, at the least, get access to an employer's health insurance plan.[2]

The U.S. Department of Agriculture's National Organic Program does include the level of a farm's biodiversity as part of its compliance assessment for certification.[3] But there are no widespread food labeling systems that pay producers premiums for reclaiming an eleven-acre marsh, putting in five miles of shelterbelts, erecting bluebird boxes, or putting in a small pond just for frogs.

It's obvious those additions to the farm help feed the couple's passion for all things wild, a passion that they've nurtured all their lives. Martin grew up on this farm, and always loved the less tame parts of it. So when he graduated from high school, he enrolled in the wildlife management

Loretta and Martin Jaus see the presence of wildlife on their dairy farm as a sign of success.

program at Central Lakes College in Brainerd, Minnesota. He met Loretta while working at a wildlife research facility in Illinois—she was doing an internship there while studying wildlife biology at the University of Wisconsin–Stevens Point. By 1980, they found themselves back on the Jaus family farm, which had been homesteaded by Martin's great-grandfather in 1877.

"I figured, well, that was a waste of my wildlife degree," quipped Loretta.

While in college, Loretta had dived deep into learning everything she could about wildlife and couldn't wait to get out in the field and do the kind of work she envisioned her education had prepared her for. She loved ornithology and all the other "ologies," but wasn't as thrilled to be required to take classes such as soil science. Loretta also vaguely remembers reading Aldo Leopold's writings and gleaning something from them about the importance of not letting immediate economic value govern which aspects of the land are retained, and which are tossed.[4]

"You know, where the first precaution of intelligent tinkering is to save every cog and wheel," she told me.

Soon after they arrived on the farm, Loretta and Martin decided they were going to produce milk in harmony with nature as much as possible. At the time, the Jauses were not familiar with the idea of practicing organic agriculture on a commercial scale, but the farm was already producing milk with few chemicals after Martin's father noticed increased abortions in the cattle and linked it to herbicide use. All the couple knew was that chemicals were not good for wildlife, and they wanted to at least attempt to make the operation somewhat friendly to the birds, mammals, and insects they loved to see on the land.

One early connection the couple made between wildlife and good conservation farming is when they planted trees as windbreaks to reduce the rampant wind erosion that can rake across this part of Minnesota's former prairie. A variety of birds, including flashy species like Baltimore orioles, responded by making the farm their summer home. It turned out the windbreak was good for the wild (birds) *and* tame (crop fields) aspects of the farm. Maybe this was a sign there was a way to connect their wildlife management background with their agricultural profession?

Thus began a series of changes to the farm that included the addition of diverse crop rotations, implementation of a managed rotational grazing system, and establishment of natural habitat. They eventually certified the land organic in 1990, and after hearing about an organic market for milk, got the same designation for their cow herd in 1994 (they now sell to Organic Valley, a farmer-owned cooperative based in La Farge, Wisconsin). The first few decades on the farm were full of trial and error. Some things, like monitoring the impacts of their practices on bird life, came easily to the trained wildlife biologists. Others, like building a diverse crop rotation without the help of chemicals, involved a steeper learning curve.

Ironically, Loretta didn't really get Leopold's concepts until she and Martin were deep into the transition of the farm to organic. She also started to appreciate the soil science she had been forced to study.

"I now understand the art of intelligent tinkering and not tinkering so intelligently," she told me during one of my visits to the farm, adding that there is no more apt place to apply such a concept than the world beneath our feet. "I began to understand the soil is where everything starts. I was now managing subterranean wildlife."

The Jauses have made use of government conservation programs to help alleviate the cost of planting trees and turning those eleven acres back into a marsh, but in general they do not get paid directly by the marketplace for establishing natural habitat on their farm. Nevertheless, they're convinced such ecologically based tweaks here and there benefit the farm's overall economic health. The Jauses' five miles of shelterbelts provide wildlife habitat, but also shelter their cows and prevent soil erosion on their crop fields. They've planted native warm-season prairie grasses in their pastures, which provide habitat for beneficial insects and help build soil. These plants also provide the kind of diversity that can help get their grazing areas through the "summer slump" that afflicts non-native, cool-season grasses. And since they are certified organic, there are no chemicals to kill off all the insects, birds, and other wildlife that can keep pest species in check. For example, the tree swallows that thrive on their pastures help control flies and other bugs.

"The more diverse the plant and animal species, the more stable, including stability for the farm," Loretta said.

They also have in place a crop rotation system that not only protects wildlife but is good for all the subsurface critters that help make good soil. The Jauses see row crops such as corn as soil depleters; small grains such as oats as relatively soil neutral; and grasses, hay, and other perennial forages as soil builders.

"So with our crop rotation overall we try to be soil neutral—some years we deplete, some years we build up," Martin explained. "We've done a lot of little things. Somehow it's all come together."

Sometimes those little things can produce big results.

A Message from the Underground

In 1989, west-central Minnesota was in the last year of a four-year drought. One of the results was a grasshopper invasion that could have come straight out of a Laura Ingalls Wilder book. One of the Jauses' neighbors had only skeletal stems and leaf veins left on his soybean plants after the insects ratchet-jawed their way through the field. When Martin and Loretta would walk into their house, fearless grasshoppers would be clinging to them while others would be creeping around the windows and doorframes in a nightmarish horde. Insecticide sprayers worked on an emergency basis day after day to save the remaining crops in the neighborhood.

The timing of the outbreak was particularly bad for small grains such as oats, which are crucial to the Jaus rotation. These plants had just started to head out, producing a succulent banquet for the voracious hoppers. After swarms passed, some small grains fields looked like they had been run through with a sickle mower. Loretta and Martin dreaded even going to the part of the farm where they had their small grains planted. But they steeled themselves for a visit, figuring that at least they could use the leftover stubble for straw.

"I still remember rounding a bend in the road and it was ragged around the edges but the rest of the field was harvestable," Loretta recalled. "We were just dumbfounded. We had no idea why that was when right across a grass buffer strip there was a neighbor's field that was pretty much decimated."

She and Martin did some ecological detective work and concluded that because their farm had more biodiversity, the grasshoppers had gone for the easier pickings in their neighbors' less varied fields. Pests love it when they happen upon one big monocultural expanse of real estate, making it possible to feed continuously without disruption. Just think what would happen if each portion of your meal was located in a different room in the house, rather than on one table. A landscape broken up into different species of crops and interspersed with perennial vegetation such as grass and trees simply makes for a more frustrating suppertime. The Jauses also wondered if maybe the grasshoppers had found something in the eleven-acre marsh that was more to their liking and were simply drawn away from their grain fields by that. The experience left them feeling positive about their decision to make biodiversity a cornerstone of their operation.

But five years later the farmers learned that the answer to this mystery had deeper roots, so to speak, than they could ever imagine. While Loretta was telling the "grasshopper story" during a bus tour of the farm, a veterinarian who works with organic farmers interjected to explain what really happened. It turns out diversity *had* saved their fields: all those rotations and other soil-building measures had cooked up such a complex, healthy biome that their fields produced grains extremely high in sugars. Sugar is a critical ingredient in making intoxicating beverages—I remember an uncle of mine talking about how back in the day he mounted heavy-duty shock absorbers on the back of his coupe so he could make money hauling *lots* of sugar for moonshiners. When the grasshoppers began feeding on the Jauses' plants, the sugars metabolized into alcohol, which proved fatal, or at least "discouraging" for the suddenly drunken invertebrates.

"Don't feel too bad about it—they died happy," Loretta recalled the veterinarian telling her and Martin.

It's a funny story to tell over a beer or coffee. But it was also an important lesson for the always observant Martin and Loretta. Don't always assume your first guess is correct and, frankly, you aren't always going to understand the nuanced connections between a farm's agronomic and ecological systems.

"It was definitely an epiphany for us," Loretta told me. "This particular example just really inspired us and set us on a new course of thinking and respect. We don't always understand why things work the way they do, but we have faith." In a sense, it's another way of saying the first lesson of intelligent tinkering is to save all the parts. You must accept that sometimes the land has answers to questions you haven't even asked.

The soil-friendly system has also paid off for the larger community beyond the farm. The Jauses are at the headwaters of the middle branch of the Rush River, which runs some twenty-three miles south before draining into the Minnesota River. The Minnesota, which eventually dumps its load into the Mississippi River in the Twin Cities, is considered one of the most polluted waterways in the Upper Midwest, due in large part to the fact that it flows through an area that's intensively farmed.[5]

One day in their kitchen, the Jauses pulled out a pair of photographs taken in the aftermath of a rainstorm and laid them on the table in front of me. One showed water flowing out of a pipe that drains a local field planted to row crops. It was saturated with chocolate-colored sediment and the volume was so great that the pipe's mouth wasn't visible. The other photo was of water leaving one of the Jaus pastures. The flow was low and clear. Martin and Loretta are low-key and humble—the opposite of show-offy. But a few years after those pictures were taken, in a don't-try-this-at-home moment, Martin allowed himself to be photographed drinking water straight out of a tile line draining one of his pastures. To get an idea of just how risky such a stunt is, consider this: in many rural parts of the Midwest, farm wells are so contaminated with nitrogen fertilizer and other agrichemicals that newborn babies must be fed bottled water purchased from town. Some dairies have been forced to drill new wells at significant cost in an attempt to gain access to water that won't contaminate the milk they produce.[6]

Working for Wildlife

Profitable milk production, healthy soil, and clean water—these are nice indicators that what the Jauses are doing here is good for them, the land, and the community. But spend any time with Loretta and Martin and it becomes clear that what really provides them the signs of success they're looking for is the wildlife. They take note of numerous mammals in their fields, including lesser-known species like meadow jumping mice. Even frogs and other amphibians, key indicators of the health of the environment, are making a comeback on their farm. And then there are the birds. The Jauses have seen the number of bird species on the farm increase from around a dozen to over two hundred over the years, including increasingly rare loggerhead shrikes and, most recently, egrets. A research team from South Dakota State University has come out to study the Jaus wetland and was excited to find clay-colored sparrows, which are increasingly hard to find in farm country.

In 2014, bird experts from the Minnesota chapter of the Audubon Society and the U.S. Fish and Wildlife Service, among others, came out on a June day to survey the farm's avian life. Kristin Hall, the conservation manager for Audubon Minnesota, said at the time a lot of her fellow birders didn't want to take part in such surveys in agricultural areas, instead opting to go to northern Minnesota, which, with its vast stands of forest, are more "birdy." In a sense, Hall felt she had drawn the short straw. But she agreed to join Tom Will, a Midwest regional bird ecologist for the U.S. Fish and Wildlife Service, on a tour of the farm.

"I thought, 'Well this will be interesting.' I hadn't really been birding in agricultural Minnesota, and I really didn't have any expectations," recalled Hall.

But she knew something was different when they approached the farm and saw the grass, brushy fence lines, bluebird boxes, and most important, the birds. Birds everywhere. It was a stark contrast to the miles of corn and soybeans she and Will had just driven through. It was as if all the birds in southern Minnesota had gotten the call that an oasis of shelter and food was present on one little dairy farm north of the town of Gibbon. Hall and Will tallied over fifty different species on the Jaus property that day, including a nesting pair of relatively rare long-eared owls.

"I think meeting Martin and Loretta changed my philosophy as to what farming should be, not necessarily what I expected it was," Hall told me later.

Will was similarly impressed. With the practiced eye of a professional ornithologist, he noted the presence of nesting habitat, insects, and general biodiversity, and knew this farm was no typical agricultural operation. But the feeling he got while spending the day roaming the Jaus land went beyond wildlife biology 101.

"As a scientist I shouldn't say this, but I think the birds know," he said. "The birds somehow can tell when someone cares. That day was one of the peak days of my life."

The birds know that wildly successful farming requires more than a pond here and a hedgerow there; it must be integrated into the daily working aspects of the operation as well. And when it is, not only the wildlife benefit—farmers can have their load lightened considerably as they go about trying to make a living on the land. Martin and Loretta made that clear one fall day when leading me and a group of natural resource professionals representing various agencies—U.S. Fish and Wildlife Service, Minnesota Department of Natural Resources, the local Soil and Water Conservation District, and the U.S. Department of Agriculture Natural Resources Conservation Service—on a tour of the farm.

They knew their audience, so the couple started the day by showing off the restored wetland-prairie area, their ponds and some wildlife habitat plantings. The agency staffers, some of whom were trained as wildlife biologists, were appropriately impressed and asked questions about what and how many species visited these areas. Then the tour moved to the pastures and crop fields that make up the operation. These areas were the heart of the farm, and served as the setting for an important message the Jauses wanted to convey: there is more than one way to skin the conservation cat.

"Normally when you think of conservation projects, you think of areas like this that are just set aside for wildlife," Martin said while gesturing in the direction of the eleven acres of prairie and wetland habitat. Then he pointed to the grazing paddock we were now standing next to. "But a well-managed pasture is also a conservation area. There are different ways a farm can help the environment."

He then walked us over to a hayfield and used a potato fork to turn over a fragrant double handful of black, black soil. It was seething with worms, bugs, roots, and the stuff of life. Given the right circumstances, soil can be so buzzing with life that it can literally generate an electrical current, and this clump was high voltage. The clean smell of actinobacteria—a sign the soil's organic matter was so healthy it was cranking out its own fertility—wafted up from Martin's hands. People crowded in to eye it and take in the sweet fragrance like it was vintage wine.

Finally, before the natural resource professionals climbed into their vehicles and headed back to their office cubicles, computers, and spread-sheets, the farmer brought his lesson home by doing what came naturally to him: sharing his joy of living on land that is home to more than a steady income.

"Every day we see something that just amazes us," he said with a smile. "One day I was making hay and I had four raptors strike mice within twenty feet of the tractor. It was two red-tails, a Swainson's, and a kestrel. A lot of people don't get to see that."

2

A Place in the Country

Improving the View in the Midst
of an Industrial Landscape

Jan Libbey and Tim Landgraf hiked through waist-high prairie to the top of a dramatic knob on their farm in north-central Iowa. As they stood among big bluestem, Indian grass, and switchgrass, corn and soybean fields flowed in every direction, the monocultural landscape broken up only by a string of wind generators to the east and a complex of confinement hog barns to the northwest. The hill Libbey and Landgraf stood on, this particular September afternoon, rose above a shorter bump in the land. The two hills, which at a distance resembled an immense sow and a piglet sleeping under a blanket, together formed a smooth, elongated ridge that ran for a few hundred yards back toward the house and outbuildings that made up the core of the farm. Geologists call such knobs in this part of the country moraines, and they are dramatic symbols of the historical glacial movements that played such a foundational role in this land's present. Just as geological activity made Texas the oligarch of oil, it played a key role in making Iowa the monarch of maize.

These ridges run roughly north-south, following the pattern of other geological formations that make up the Des Moines Lobe, a tongue of immensely rich soil that was left behind when the Wisconsin glaciation pulled back its icy blanket some twelve thousand to thirteen thousand years ago. The melting glacier left behind a poorly drained landscape of pebbly deposits, as well as clay and peat from glacial lakes. It made for

When Jan Libbey and Tim Landgraf first bought their land, they viewed it through the lens of traditional environmentalism: farming and nature did not mix.

some of the best, flattest farmland in the world, but here and there left what's been called a "knob and kettle" topography—small lakes, potholes, and, occasionally, bumps on the land.[1] Thanks to one of those larger bumps, the state capitol building in Des Moines sits on high ground in the south-central part of the state.

A geological anomaly like a moraine on an otherwise flat landscape catches the eye. Devils Tower might not be so impressive if it was plunked down in the middle of the Rockies, but a rise of land a few hundred feet high in the middle of Iowa can be jarring. If you're a corn farmer whose goal is to drag a forty-eight-row planter across a tabletop landscape at ten miles per hour, that change in elevation is to be avoided. But if you're someone who's looking for a nice view out in the country, it's quite welcome indeed. In a part of north-central Iowa famous for its flatness, the couple had somehow found elevation.

"These hills are part of the reason we moved here. We weren't going to farm—we just wanted a big acreage," said Landgraf, adding that the previous owners "were just tickled pink to sell it to us, because we took all the rough ground."

When the couple bought this land in the heart of Iowa's row crop country in 1989, they viewed it through the "lens of the traditional environmentalist," recalled Libbey. But a lot had changed in the past quarter-century. That fifty-five acres of land they originally bought has expanded to 132, and it is now the source of their livelihood, not just a respite from the industrialized world. In 1996 Libbey and Landgraf launched One Step at a Time Gardens, a Community Supported Agriculture (CSA) produce operation. They started out with six members and over the years steadily grew it to the point where in 2002 Tim could quit his town job as an engineer. When I first visited the farm in 2009, Libbey and Landgraf had 150 farm shareholders in nearby towns as well as Des Moines, one hundred miles to the south, and were producing vegetables on eight acres of gardens. By the end of 2016, the couple had downsized considerably—they had pared their vegetable acres down to four and had fifty shareholders—deliveries were no longer being made to Des Moines. Besides the CSA market, they were selling wholesale in the region, as well as marketing pasture-raised meat chickens direct to consumers. A portion of their wholesale business included membership in North Iowa Fresh, a start-up food hub that sold to grocers and restaurants in northern Iowa.

Despite these changes, Landgraf and Libbey have remained committed to retaining the land's natural habitat. It's a balancing act, one they think is made easier thanks to the model they use to market their food. They got into farming to have a more legitimate voice in the debate over the future of rural communities in the area. Now the question is how can such a wildly successful farm introduce the rest of the community to a different idea of treating the land?

Fighting for the Land

Jan Libbey grew up in Des Moines and has a fisheries and wildlife degree from Iowa State University, where she met Tim, who has a degree in metallurgical engineering and was raised on a diversified farm. They both have a love of the outdoors. While working as a county naturalist in north-central Iowa for five years, Jan was able to put into practice her passion for connecting people to the land utilizing environmental education. So it seemed like a natural step for the couple to move out of town and get onto some land. What particularly attracted them to the hilly parcel they ended up on was that it was right across the road from the 490-acre East Twin Lake Wildlife Management Area, a public gem consisting of a glacial lake, wetlands, and forested land. They rented out the farm's crop acres and began raising a family—they have two grown children, Jess and Andrew—while Landgraf continued his career as an engineer.

But soon after moving to the farm, a neighbor proposed building a large industrialized hog operation in the neighborhood. This dragged Libbey and Landgraf into the heart of the factory farm hog wars that had begun in the Midwest a few years before. As reports of water and air pollution caused by liquid manure contamination from large concentrated animal feeding operations—also known as CAFOs—proliferated, it was beginning to look like a healthy environment and "modern" farming were not compatible. This created significant tensions in rural communities.[2]

"During my early work in environmental education, it was more common to talk about agriculture in terms of its environmental detriments. It put farmers and environmentalists at loggerheads," said Libbey, bumping her fists together to illustrate her point.

The battle they and their neighbors waged went all the way to the Iowa Supreme Court, which ruled that local governments such as counties could not control the siting of factory livestock operations. The result of that decision is evident when one visits One Step at a Time Gardens on a summer day—at times the pungent odor of the hog CAFO a half-mile away wafts over the farmstead. After the court decision, Libbey and Landgraf sought out a different avenue for dealing with the situation:

they more than doubled the acres they owned to prevent future CAFOs from moving closer.

"We found out we weren't warriors," said Libbey of their foray into the factory farm wars. "The lobbying work we got involved with seemed to be more about putting up stop signs and less about offering new directions. As landowners, we began to see the need to open dialogue, not shut it down."

They discovered that in their immediate community, there simply was no farming model that offered a viable alternative to large-scale industrialized agriculture. And Libbey was aware of research that came out in the early 1990s showing that even when environmental educators get people out on the land for hikes and other activities, it doesn't necessarily change behavior. Maybe Libbey and Landgraf weren't frontline eco-warriors, but that didn't mean they couldn't participate in the battle in other ways. So they started looking at how they could model a more sustainable kind of agriculture in their community. As part of their research, one year the couple traveled to southwestern Wisconsin and attended the Organic Farming Conference that's sponsored annually by the Midwest Organic and Sustainable Agriculture Education Service (MOSES). There they met farmers who were proving food production and environmental health weren't mutually exclusive.

The Community Supported Agriculture model appealed to them, partly because they already knew how to raise a big garden. But the way CSA farming connects producers and consumers was also attractive to the couple. It consists of people buying a share in a farm before the growing season. In return, the operation provides regular deliveries of sustainably grown food. Most CSA farms provide produce, but an increasing number are also offering members other products from the land, such as meat, cheese, eggs, and even cut flowers. CSA farming is considered by conscientious eaters to be one of the ultimate methods for knowing intimately the source of one's food and thus having a say in how and by whom it's produced.

"I think it was the depth of the model that appealed to me," Libbey told me. "It provides for an extension of that naturalist education."

Alleys and Ecosystems

Over time, the environmentalist and farmer worlds have melded on the farm, something that is possible when you're producing food in a manner that does not rely on thousands of acres of monocrops. Libbey's goal is to make the farm into one that reconnects food systems and ecosystems on a daily basis. For example, she and Landgraf showed me a part of their gardens that is planted between two sets of shelterbelts—a part of the farm they call the "Alley." It builds on some soil conservation tree plantings of poplar and honeysuckle that were already part of the farm back in 1989. The farmers have since added oak, walnut, and ninebark shrubs to the shelterbelt. Besides the soil conservation benefits, the thicker plantings provide a microclimate that alleviates the kind of weather extremes that can make vegetable farming tricky in the Upper Midwest.

At one end of the gardens were the farm's main source of fertility: pastured chickens. They are part of an intricate rotation system where the farmers take two of the fields out of production every year and seed down red clover. They then run the chickens over the red clover, moving the pens daily. The next year the chickened land, rich in fertility and tilth, goes back into food production. The chickens aren't just a source of fertility—the family sold around nine hundred a year at the peak of their production.

After checking on the chickens, Libbey and Landgraf left the shelter of the alley gardens and visited more gardens that are planted on the exposed ridge of the lower part of the moraine. The lack of tree cover was noticeable, as stiff winds swept the ridge. The farmers had also planted oak, American cranberry, and spruce along the sides of the ridge-top gardens, a future source of protection from the wind.

"We wished we had planted these earlier," Libbey lamented. But then, they didn't know when they first moved here they were going to have six acres of garden to protect. North of the ridge-top gardens was a fourteen-acre wetland restoration that was established in 2001. Native grasses were planted in the upland of the restoration, and cattails poked up in the open water. On the other side of the ridge was a fifteen-acre wetland that was established in 2008. After checking out the wetlands, Landgraf and Libbey waded through the eight acres of prairie that was restored on the ridge soon after they moved to the farm.

All of this has made for a farm that has a nice mix of cultivated and wild land—the prairies, tree plantings, and wetlands seem to wrap around the gardens and the row crop acres they rent out to a neighbor. Spend any time here and it will become clear that although they are now making a living on this land, these natural pockets are still key to Libbey and Landgraf's quality of life.

"It's a lot of hard work," said Libbey of producing food on a weekly basis for CSA shareholders. "So, within this hard work you always need this respite that's kind of close to our core in the first place. This is what drew us here."

Posted on the wall of a garage that's been converted into a vegetable packing shed was a listing of forty bird species they've spotted on the farm. Geese, ducks, herons, and swans, as well as deer in the wintertime, utilize the restored wetlands. Landgraf described a recent day in the fields that was a wealth of wildness: "So first thing there was a big flock of probably thirty or forty pelicans. When they're flying that's when we'll see them—this big salt and pepper circling. And then we had fifty or sixty Canada geese." The farmers also talk with delight (and a little pride) about the migrating monarch butterflies that congregate by the hundreds in the alley garden area and the upland chorus of frogs that hang around the places where the chickens graze.

A Public Good

But unlike the East Twin Lake Wildlife Management Area across the road, One Step at a Time is not devoted exclusively to serving as a nature preserve—it must pay its own way. In the early years of their land tenure, Landgraf and Libbey's management decisions were driven primarily by aesthetics; these days taking acres out of production for prairie or wetlands is something the farmers think twice about.

"I think the key is trying to have that planting diversity and trying to figure out how you can intersperse what you want to produce with wild areas," said Landgraf. "Because at the end of the day we have to produce product to make some money so we can stay here. You know, we're not just doing this for kicks."

Government programs have helped. For example, the wetlands and their adjoining uplands are a result of the Wetlands Reserve Program, a United States Department of Agriculture Natural Resources Conservation Service (NRCS) program that provides cost-share money for establishment of wetlands, as well as a regular per-acre payment that is comparable to what Libbey and Landgraf would receive if they were renting the land out to a crop farmer. Some of their natural areas are set aside in the Conservation Reserve Program, which also provides annual "rental" payments.

The NRCS also provided cost-share funds to help pay for the trees. In addition, the Hancock County Conservation Board has provided a tree planter, seed drill, and prairie burning assistance to help establish and maintain the farm's habitat restorations. The farmers have also taken advantage of state and federal assistance to move their operation in another sustainable direction: energy independence. The couple has erected an impressive array of solar panels on a hillside near their specially designed energy-efficient house, which was built after the original one burned down. The analytical engineer in Landgraf emerged as he excitedly described the analysis that went into choosing solar over wind energy. It turns out their peak demand is in the spring and summer, when the walk-in cooler is working overtime to keep produce fresh, and they are burning a lot of electricity to freeze broilers. Solar panels churn out the most electricity right at that time of the year. Wind turbines, on the other hand, are higher maintenance and at a low ebb power-producing-wise during the summer in that part of Iowa. On one visit to the farm, I walked over to examine the solar array, which was soaking up a summer day; at times the panels produce more electricity than the farm can utilize. A few miles away a collection of giant white wind turbines stood baking in the sun among thousands of acres of corn and soybeans, their rotors as still as statues.

The practical decision-making that went into going solar is typical for the way Libbey and Landgraf approach management of their farm. The environmentalists in them are interested in adopting sustainable energy alternatives and establishing wildlife habitat; the practical farmer side requires some numbers to be crunched first. Just because it's better for the environment doesn't mean a new addition will automatically get the nod. Environmental sustainability and economic sustainability must go hand-in-hand.

That said, the farmers are the first to admit that even when they've used government conservation programs on the farm, it's not always as well planned out as it might appear. For example, the two wetlands they've established were mostly out of desperation—the farmer they are renting some of their land to was tired of getting stuck in former prairie potholes.

"We finally said forget it—it wants to be a wetland," Libbey said.

In some ways, Landgraf and Libbey are using government conservation programs to buy time for their land and figure out its long-term future. For example, the fifteen-year Wetlands Reserve Program contracts help alleviate the financial risk of keeping that land in natural habitat for the time being. But the clock is ticking, and it's not clear whether such contracts will be available for renewal in the future.

"The payment is pretty comparable to what you're getting for cash rent, so economically that solves that issue for fifteen years," said Landgraf. "But then at the end of the fifteen years you've got to say okay, now what? Because we can't afford to keep it permanently out. We're going to have to do something."

Maybe by the time the contract runs out, they will have figured out how to make the ecological services provided by such habitat—cleaner water, a home for wildlife, and so on—pay off, either through the marketplace or via an acknowledgement that natural farm habitat is a public good society feels is worth supporting. For example, while attending the MOSES Organic Conference several years ago, they ran into Eric Mader, who worked for the Xerces Society for Invertebrate Conservation. Entomologists, farmers, and beekeepers are becoming increasingly alarmed at the decline in domesticated pollinators such as honeybees, as well as their wild cousins such as bumblebees.[3] This is one of those ecological crises that affects humans directly—every third bite of food is directly or indirectly the result of the work of pollinators. Produce farms especially are heavily reliant on pollination services.[4]

Beekeepers are reporting cases where so-called colony collapse disorder is wiping out entire hives with no warning.[5] In early 2017, the rusty-patched bumblebee had the dubious distinction of becoming the first bee in the continental U.S. to be put on the Endangered Species list. It's estimated that this species alone has declined in nearly 90 percent of its range over the past two decades.[6] The problems faced by pollinating insects have been attributed to, among other things, pesticide poisoning, imported diseases, climate change, and lack of foraging resources caused by the plowing up and paving over of the landscape. University of Minnesota bee expert Marla Spivak told me there's likely no single cause—it's probably a combination of factors. The issue that overshadows every other threat to wild and domesticated pollinators is lack of natural habitat to forage on and live in. Diverse landscapes can go a long way toward making beneficial insects more resilient in the face of disease, toxic chemicals, and general stress, say entomologists and ecologists. And this is where diverse, sustainable farms can play a role.

Research shows that farms with woods, meadows, and other natural areas growing flowering plants have a larger number of insect pollinators. But monocultures of corn are deserts to such insects.[7] Beekeepers often get panicked calls from fruit and vegetable producers who are trying to raise melons and other pollinator-dependent crops in the midst of corn country. In addition, heavy tillage disrupts wild bee habitat— two-thirds of native bees nest underground. And according to research conducted by entomologists at Pennsylvania State University and North Carolina State University, plants grown in healthier soil—in this case soil treated with earthworm compost to increase biological activity—are much more attractive and nutritious for pollinators when compared to their chemically treated counterparts.[8] That's one more reason we should be excited about efforts to improve soil health on farms (see chapters 5 and 6).

At the Organic Conference, Mader talked to Libbey and Landgraf about the role diverse operations like theirs could play in providing habitat for wild pollinators such as bumblebees. The conversation struck a chord with the farmers, who rely heavily on pollinators for their gardens. At the time, they had recently lost access to some honeybee hives, and were wondering where they would get a consistent source of pollination services. Mader made it clear that all of the native prairie plants they have interspersed around their gardens provide prime habitat for pollinators. Indeed, research done by Iowa State University showed One Step at a Time is home to a good diversity of pollinators, particularly various species of squash bees. And those pollinators can benefit not just their farm, but agricultural and natural plantings in general—providing an ecological service to the community at large.

When I returned to the farm seven years after my first visit, Libbey and Landgraf had just got done enrolling four acres into a USDA program that provides landowners funds to establish and maintain pollinator habitat. The plan was to plant eight different grasses and thirty-three different forbs.

"You know, maybe the benefit I get from that piece of ground is that it is an incubator for pollinators that make food production possible in my gardens," said Landgraf. "What other benefits are we getting in terms of pollinators, in terms of beneficial insects, reduced insect pressure, that we really don't know about? Perhaps it's enough to get you rethinking how you view a piece of ground that you think you need to be farming."

The four acres of pollinator habitat is being borrowed from a fifty-five-acre parcel Libbey and Landgraf are cash renting to a local crop farmer. Odd corners that were hard to farm anyway are being turned over to the perennial plantings.

Libbey, for her part, loves how issues like the need for pollinator habitat can bridge the divide between the environmentalist and farmer worlds that exist within the community, as well as within herself. "Gosh, that environmentalist bent is integrating itself into the ag bent," she said.

Takes a Community

The hundreds of thousands of acres of monocropped fields that flow in every direction from One Step at a Time are a constant reminder that no matter what innovative changes the farmers make on their own land, a whole lot of status quo is out there. Libbey and Landgraf feel members of the community at large—farmers and nonfarmers alike—must show they are willing to support farms as natural habitats if operations like theirs are to be more than the exception. One way the general public can do that is to consume food produced on sustainable farms that are benefiting the local environment.

Libbey worked for a couple of years on a project in Wright County to get local food to low-income people. She helped build up the North Iowa Farmers' Market in Mason City, and assisted in coordinating a Leopold Center for Sustainable Agriculture-funded project that focused on local food as economic development.

"If we can take that regional approach, then we can start putting in place some initiatives that begin to trickle down and have a more local impact," said Libbey. "Because the farm and food connections are still happening on a very local basis."

"Hopefully our farm members are making the connection that agriculture is not just something that's done out there," added Landgraf, gesturing to the surrounding countryside.

He and Libbey make that connection through the food, of course, but also via a weekly newsletter that comes with the shares during the growing season. Besides recipes, a listing of what's in the share, and an update on the farm, the *Weekly Note* also includes ways—book suggestions, meeting notices, brief notes—of connecting people to the larger issues affecting agriculture, the land and communities.

But there's still some work to do before One Step at a Time becomes a "social change agent," as Libbey puts it. She said members tell them the main reasons they belong to the farm are that it's a source of healthy food and they like the idea of supporting a local farm. They also say they belong for what they term "environmental reasons." But when

Libbey presses them further on that particular point, it becomes clear these members are mostly concerned about specific practices that keep chemicals out of their food, rather than the "big picture sustainability" of family farms and rural landscapes.

Changing minds in the agricultural community may be even tougher. The couple is active with Practical Farmers of Iowa (PFI), which, among other things, conducts on-farm research and field days related to sustainable agriculture. Landgraf sees even conventional farmers showing an interest in trying sustainable production systems as they grapple with problems like soil erosion and pest outbreaks. He's also excited that the environmental community is starting to see that food production and ecological health can go hand-in-hand. Groups such as the Xerces Society and the Iowa Natural Heritage Foundation have co-sponsored PFI fields days at One Step at a Time. Landgraf pointed out that such field days cover more than what is growing on the tillable acres.

"At most PFI field days you focus on what's in the fields," he told me. "At ours we focus mostly on what is *around* the fields."

The Glacial Grind

On a recent visit to the farm, I got a sense of what Landgraf was talking about. Around thirty of us had gathered at One Step at a Time Gardens on a sunny Sunday in mid-August for a PFI field day. The focus was on the farm's work with pollinators, as well as its research related to nineteen heritage tomato varieties and how to do enterprise analysis for a working farm.

After introductions beneath a giant maple tree in front of Jan and Tim's house, we walked across the farmyard to two low-tech greenhouses called high tunnels. With their plastic sheeting stretched over hooped metal tubing, high tunnels resemble miniature Quonset huts, and in recent years they've become a popular way for midwestern vegetable producers to extend the growing season—they trap a surprisingly large amount of the sun's rays late into the fall and early in the spring. A few yards from the tunnels were two abandoned bear-cage corn cribs and a steel-sided grain bin. One high tunnel was fairly packed with tomatoes; clusters of fat fruit were strung up to make efficient use of the space. The other high tunnel was growing fall greens like lettuce and kale, as well as peppers. Harvested onions had been hung up to dry along the sides of one high tunnel.

The rich black soil in the vegetable plots downhill from the high tunnels was covered with a variety of vegetables, as well as a sign of one disadvantage of farming within a rifle shot of a wildlife management area. Despite the presence of a tall electric fencing system, the dainty hoofprints of deer were everywhere. "It's really a deterrent—it's not 100 percent," Landgraf said of the fencing, adding that deer have even jumped through the side panels of the high tunnels to feed on plants.

While standing in the vegetable plots, Libbey described the importance of developing good financial enterprise analyses for small farms like theirs. She held up two resources they rely on—*Fearless Farm Finances* and the *Organic Farmer's Business Handbook*—and described how One Step at a Time used enterprise analysis to determine, for example, that a barrel washer for produce was worth the investment because of its efficiency. They've also used enterprise analysis to determine how to charge a profitable price for their products. It's clear Libbey takes good business planning seriously—people call and e-mail her for advice on such matters. But suddenly, the farmer cut herself off in mid-sentence.

"Pelicans! Awesome! I've always said I'd like to be reincarnated as a pelican," she shouted as her head jerked skyward.

A couple dozen other heads followed Libbey's lead. For a moment, farm finances, the predation of pesky deer, and what to charge for a dozen ripe tomatoes faded into the background as half a dozen black-and-white pelicans winged over us in elegant slow motion.

Later, while thinking about how a few aquatic birds had forced us all to live in the moment, I recalled a discussion I had years before with Libbey and Landgraf and their daughter Jess while we were walking the farm. It was about how deep, long-lasting change on many levels—farm, community, regional, and even national—won't occur overnight. At one point Jess blurted out the word "glacial." She was not referring to the farm's geological history. After all, it took Libbey and Landgraf a few years to accept that agrarianism and environmentalism could preside on the same piece of real estate. Once they came to that realization, they struck upon a way of producing food that made room for bees, birds, and yes, even deer.

Libbey told me she and Landgraf—they are in their midfifties—are starting to accept the idea that One Step at a Time Gardens may not exist in the future as a working farm (life has taken each of their children's paths in directions that don't include farming). Now that they are beginning to wind down the business end of the operation, Tim and Jan inevitably wonder if they have had enough of an impact on the wider community to ensure people will always get the opportunity to take in a little wild drama while going about the daily business of making a living in agriculture. Ideally, an oasis like One Step at a Time will find a way to exist as a physical entity. But when one considers the sometimes overwhelming expanse of history, getting too attached to a physical piece of real estate may be naïve and a bit short-sighted. Maybe what we should be hanging onto, and promoting whenever possible, is the *idea* that with a little creativity and luck (and a lot of work), a wildly successful farm can rise within an industrialized landscape. Rocks, ice, soil, and even international grain markets come and go—ideas remain embedded in the collective human seedbank.

"Sometimes you get so wrapped up in the work that you want change to happen now," Libbey told me once. "You have to have some patience and understand our work is only going to be a piece of it. Sometimes you need to sit back and feel good about the here and now."

3

Blurring the Boundaries

Community Conservation and
the Power of a Common Goal

In the years I've spent interviewing and writing about farmers, it can seem at times that the ones seeking out alternative methods of making a living on the land are a bit of a lonely bunch. Just traveling to these farms can give one the sense of ecological isolation: a drive through thousands of mind-numbing acres of a corn and soybean duo-culture before coming upon, for example, lush, rotationally grazed pastures, some windbreaks, and even a slough or wetland. Maybe you start to see a few more meadowlarks and a red-tailed hawk or two. Islands of any type can be beautiful, but they can also be cut off from the rest of the world.

And that isolation isn't just geographic. Many ecological agrarians express concerns about being isolated professionally, or even socially. Even at a time when the word "sustainable" has become mainstream, wildly successful farmers are still often looked upon with suspicion by their neighbors and the local agribusiness community. And if they have the nerve to actually host field days for birders, government natural resource professionals, and other "tree-hugger" types, well, they may as well be dancing around a Maypole on Planet Greenpeace.

But there's nothing like a common threat to create a little community among people who usually don't see eye-to-eye on various issues. For example, these days, anyone who cares about having grass and other perennial vegetation growing on our agricultural landscape should be alarmed at the massive conversion of such habitats to annual row crops. It's become clear in recent years that even land retired and planted to grass through the federal government's Conservation Reserve Program (CRP) isn't safe from the plow—when corn prices skyrocketed to record levels in the mid-2010s, many landowners didn't renew their CRP contracts, or even paid a penalty to opt out of the contracts early, so they could plow up the sod and cash in on the red-hot commodity market.[1]

"Dirt farmers going fencerow-to-fencerow, I hate to see that," a cattle producer told me as he stood in one of his pastures on an early fall day. "I lost two hundred acres of pasture I had rented for twenty-five years to the plow when we got seven dollar corn. Three miles south of here it was all in CRP. Now it's all crops. There's a reason it was in CRP."

CRP is a taxpayer-funded program. Ironically, the loss of pasture and other grasslands is due in large part to other aspects of government policy. The *Proceedings of the National Academy of Sciences* reports that between 2006 and 2011, 1.3 million acres of grasslands were converted to crops in Minnesota, Iowa, North Dakota, South Dakota, and Nebraska. Such conversion rates haven't been seen since the 1920s and 1930s, and anyone familiar with the Dust Bowl era knows what resulted from that.[2]

A concerted effort on the part of the federal government, along with its counterparts on the state and local level, helped our rural communities recover from the Dirty Thirties. That's why it's particularly frustrating to see the role public policy is playing in the recent loss of perennial vegetation and all the ecosystem services it provides. For example, there are strong indications that the conversion of corn to ethanol, a process that has benefited from numerous state and federal government subsidies, plays a major role in putting more grass under the plow. One significant federal government program is the Renewable Fuel Standard, which mandates that a certain percentage of gasoline sold in the U.S. contains ethanol. This rule has been credited with accelerating the demand for corn. One study showed that during initial implementation of a particular version of that law from 2008 to 2012, nearly 4.2 million acres of land in the U.S. was converted to crops within one hundred miles of ethanol refineries. Of that, 3.6 million acres had been in grassland. Overall, the rate of conversion to cropland has increased the closer the land is to an ethanol refinery.[3]

A University of Wisconsin study published in 2015 used high-resolution satellite data to track how much new cropland we gained in the U.S. between 2008 and 2012. Among other things, the researchers wanted to determine if demand for crops such as corn fueled the conversion of previously uncultivated acres. They found that nationwide, over 7.3 million acres of previously uncultivated land was converted to crops during the study period. Seventy-seven percent of that new cropland came at the expense of grassland—native prairie, pasture, and hay ground. Corn was the number one choice for planting on newly broken ground, followed by wheat and soybeans.[4]

Ironically, one of the arguments for blending ethanol into gasoline is that it will reduce greenhouse gas emissions. But studies like these call that into question, considering all of the carbon released when land in perennial vegetation is plowed up and converted to an annual row crop. The 2015 University of Wisconsin study found that carbon emissions produced from corn and soybeans planted on recently tilled land would be equivalent to a year's carbon dioxide release from thirty-four coal-fired power plants, or twenty-eight million cars. Given that major impact, the researchers concluded that stricter enforcement of rules around expanding cropland for biofuels production are needed.[5]

"We could be, in a sense, plowing up prairies with each mile we drive," Tyler Lark, one of the study's authors, quipped in a press release.[6]

In west-central Minnesota, I found a group of dedicated livestock producers who are scrambling to save every last bit of grazing land they can in the face of the plow and benign neglect. This common struggle has created allies among grass-based farmers who, just by the fact that they want to preserve perennial vegetation, are seen as out of step with the norm of raising nothing but annual row crops such as corn and soybeans. As it happens, the environmental community in this region also wants to save grass, albeit for different reasons. This common goal serves as the basis for an initiative that's helping the parties involved see beyond old cultural barriers and to, in fact, break down physical ones: such as the fence line dividing a working farm from a wildlife refuge.

A Disturbing Development

As a scientist with The Nature Conservancy based in a midwestern state, Steve Chaplin thinks a lot about the impact agriculture has on ecological treasures such as native tallgrass prairie. "Other than plowing, grazing has probably been responsible for the degradation of more prairie than any other source," Chaplin, who is in the Conservancy's Minnesota field office, once told me as we stood in a nature preserve owned by the environmental organization in west-central Minnesota's Pope County. No surprises there. But less expected was Chaplin's next words: "We would like to see grazing on a large scale, which would mean grazing across public-private property lines. To a lot of conservationists, it is probably surprising that we need more people, rather than fewer people, to improve the landscape."

More farmers, and by extension, the cattle they manage, means more disturbance, and that's a good thing. It turns out native prairies, other grass-based habitats, and even wetlands need a little disruption of growth patterns if they are to remain healthy ecosystems, rather than scrubby patches of land covered by red cedar and other invasives. That's why in this part of Minnesota, Chaplin and other natural resource experts are welcoming cattle onto lands that were once off-limits to livestock: preserves, wildlife refuges, and other natural tracts of real estate.

Public agencies and private conservation groups are fast realizing that buying up land and putting up "Nature Preserve" signs won't secure the long-term sustainability of that habitat—it needs active management, the kind that toes the line between stressing the environment and allowing it to recover. It turns out when cattle are used to provide that well-balanced mix, the result can be a healthier, more diverse habitat, as well as an extra incentive for farmers to keep livestock as a key part of their enterprises. Although not a perfect replacement for the tens of millions of bison that used to roam the landscape in a large-scale rotation that involved a years-long balance of disturbance and rest, controlled herds of cattle are the closest proxy we have right now.

"We need to keep cowmen on the ground," J. B. Bright, a U.S. Fish and Wildlife Service refuge specialist who works with livestock farmers in western Minnesota, told me. "The local economies are stronger and the perennial plant systems are stronger."

In the Midwest, the return of cattle to prairies and other natural areas is a relatively recent phenomenon. Grazing of public lands has a long history out West, where large herds of cattle have been allowed to roam at will on public rangeland during the entire growing season, often with little or no controls. In many cases, the result has been decimated grasslands and destruction of riparian areas and woodlands, resulting in compromised wildlife habitat, erosion, and polluted water. Such an unfettered system of land use has given cattle producers a major black eye among environmentalists in the West.

In these circumstances, banning livestock from natural areas and refuges would appear to be a no-brainer. But such a rigid line in the grass can also result in lands that suffer from severe benign neglect.

Depending on the situation, grasslands require a major disturbance at least every five to ten years, something bison and wildfires provided in days gone by.[7] Natural resource experts purposely burn off grasslands to keep woody invasives at bay and recharge green growth. But managing a burn can be expensive and it requires optimal weather conditions; there is also a growing concern about how burning puts more carbon in the atmosphere at a time when we need it to be sequestered in our plants and soil. As a result, refuge managers concede they are woefully behind on controlling invasives, and they are watching with alarm as pastures purchased from farmers become inundated with cedar, Siberian elm, Russian olive, and red-osier dogwood within four or five years.

Fortunately, innovations in grass-based livestock production offer a prime opportunity to bring back the kind of flash disturbances that haven't been around since the time bison ruled the prairie. More livestock producers are starting to see the benefits of moving cattle frequently through numerous paddocks, rather than keeping them on the same pasture all season long, where it becomes overgrazed. This system can extend the grass season, cut costs, and in general produce more profits. Advances in watering systems and lightweight moveable electric fencing have made rotational grazing even more viable. Innovations in portable fencing are no small matter and have a direct connection to getting and maintaining more perenniality on the landscape. Often, the removal of fences is the first sign that a farm has been fully converted to row crop production and that livestock, and thus pasture, are no longer part of the picture. Corn and soybeans tend not to wander across property lines, and wire and posts just get in the way of large field equipment. With lightweight, moveable fencing, livestock can once again be part of a farm's future.

This type of grazing system fits well with what refuge managers are looking for: short-term impact (a few weeks) and long-term rest (a year or more), something grassland experts call "conservation grazing."

"The key is to hit it and rest it," Minnesota Department of Natural Resources prairie habitat ecologist Greg Hoch told me. "That's how these prairies evolved with the bison. I'm 100 percent convinced that if we do grazing right, grassland diversity will increase."

Rangeland science backs up Hoch's contention. Studies in numerous states show that conservation grazing can as much as double plant diversity in an area—it not only prevents overgrazing of grasses and forbs, but the cattle's manure and urine help recharge the soil's biology.[8] Hoch and other habitat experts working in western Minnesota have observed how grazing has increased native plant communities by knocking back invasive cool-season plants such as Kentucky bluegrass and smooth bromegrass. Such invasives tend to blanket the land with a homogeneous cover, which limits the diversity wildlife such as deer, waterfowl, shorebirds, and grassland songbirds prefer. Such grasses also tend to go dormant in hot weather and provide limited habitat and foraging areas for pollinators. Cattle are also being used to thin out cattails and reed-canary grass around wetlands, providing the open areas many waterfowl prefer when keeping a lookout for predators. And controlled grazing of riparian areas is proving to be an effective way to stabilize shorelines along waterways.[9]

The science has become so convincing that conservation groups such as The Nature Conservancy and the National Audubon Society have modified their once decidedly negative view of cattle and now see them as an effective habitat management tool, given the right circumstances. However, it's not likely we will see a situation where livestock are swarming over all of our public lands. Of the fifty thousand acres the Fish and Wildlife Service manages in the Morris District in western Minnesota, around five thousand acres were being grazed by 35 different producers as of 2016. The Nature Conservancy grazes less than 15 percent of the 63,500 acres it owns in Minnesota.[10]

Nevertheless, conservation grazing is seen as a potentially key tool in targeted areas. The Minnesota Prairie Conservation Plan, which was developed by ten conservation agencies and organizations, provides a blueprint for how to save and manage a resource that once covered eighteen million acres of the state, but is now down to 235,000 acres and shrinking fast. The authors of the report identified conservation grazing as a major method for preserving and managing grasslands.[11]

The Prairie Conservation Plan highlights a shared threat livestock farmers and conservationists face: the plowing up of grass to make way for more corn and soybeans. It should be kept in mind that although wildlife managers and farmers share a common desire to save grass, they can still differ widely on what that resource should ultimately produce. The farmer wants feed and the natural resource manager wants a diversity of habitat. Livestock producers usually pay a fee to graze refuges and other natural areas, but that doesn't give them carte blanche—the refuge manager's goal of protecting the resource takes precedence over profits.

But when a balance can be struck, it's a good way to manage an important resource on multiple levels, said Dan Jenniges, who has a cow-calf operation near The Nature Conservancy's Pope County preserve. Jenniges, who has been grazing land controlled by the Fish and Wildlife Service and the Department of Natural Resources for several years, said the grazing schedule and intensity can vary from year to year, depending on what an agency's objectives are for a particular piece of land. Sometimes his cattle are brought in during the spring to knock back cool-season grasses like brome and bluegrass just as they're starting growth; other times a fall grazing is called for, to stymie the same grasses as they are coming out of summer dormancy.

Some of Jenniges's land is adjacent to refuge land, making grazing the public areas convenient; in other cases he has to transport the cattle several miles for a grazing season that may only last around a month. That can be a hassle, but it allows him to give his own pastures a rest and break up pest cycles while contributing to the health of the overall landscape.

"We aren't renting the grassland—we're managing it," Jenniges told me. "When you're grazing that public land, you're able to take pressure off your own lands, so in general all the grasslands become better, whether it's for the grass or the wildlife. Without livestock, there is no reason for a community to have grass."

More Fences, More Neighbors

It all starts with the fences. Once a field goes into corn and soybeans, the landowner no longer needs fencing to keep livestock in. And once they're ripped out, it's been made clear that livestock will not be coming back to that land in the foreseeable future. Jenniges knows fences. It seems like whenever I call him on his cell phone I catch him making or repairing fencing for his cattle and sheep.

"Yeah, it's job security, it's never-ending," he told me during one of those phone calls. But that fencing is an indicator that he's committed to raising livestock on—and this is important—grass. After all, livestock can be raised in large confinement facilities, with the feed being hauled in and the manure hauled out as a waste product. But Jenniges is a "grazier," a farmer who has centered his livestock operation around having the animals harvest their own feed in the form of perennial grasses and forbs. Numerous livestock farmers dabble with allowing their animals to get some of their nutrition directly from the land when it's convenient—when they have some corn stubble or a cover crop that can be grazed for a few weeks, for example. But graziers prioritize integrating livestock and a perennial landscape. Such a system has an economic impetus, given that it saves the tremendous expense of investing in the equipment required to raise, harvest, and process animal feed. But for Jenniges, it's not all dollars and cents; he can wax downright poetic about the beauty of a well-cared-for pasture.

He showed me a lot of fences one July evening as we took a tour of the many pastures—rented and owned—he has spread throughout the neighborhood. Jenniges explained that his father was a grazier as well. In fact, he had moved to this area from southern Minnesota decades ago to escape crop farming. So when the younger Jenniges decided to go into farming in the 1970s, raising cattle and sheep on pasture seemed like the way to go. In fact, even before he graduated from high school he was raising sheep on a tree farm a local family owned; they needed the animals to thin the invasive tree species. Some four decades later, he still grazes that land. Over the years, Jenniges, along with his wife Linda, have grown their operation to include a cow-calf herd of black Angus and Simmental, as well as the sheep. They raise some corn and soybeans, but grazing livestock remains the focus.

Jenniges is an interesting mix of someone who is passionate about the future of his rural community and a bit of a lone wolf who literally does not care what others think. For example, he has gone out of his way to help beginning farmers in the neighborhood, and gets genuinely sad when talking about how full the school bus used to be when it rolled by his farm. On the other hand, he's not afraid to go completely against the grain—or whatever plant is involved. He can get pretty worked up about how he feels urban environmentalists are pushing livestock farmers out of business as a result of what he sees as overreaching regulations. I've also seen him brandish a butterfly net among naturalists and scientists at a local nature walk. To the chagrin of corn farmers, Jenniges makes a convincing argument that the "weed" quack grass is the perfect forage and shouldn't be sprayed out of existence. He once baled up cattails by driving his equipment on the ice in a marshy area. He ground them up and mixed them with distillers' grains, beet shavings, and grass before feeding them to his cattle—their hides became coated with cattail duff during chore time. "I'm not smart enough to worry about taking a chance. I'll try anything once," Jenniges told me, adding with a lopsided grin, "You shouldn't avoid having your neighbors thinking you're crazy—just don't remove all doubt."

"What's he been smoking?" his wife Linda recalled the neighbors saying after one of Dan's very public experiments involving cover crops and corn. It turns out it was Doral cigarettes, one after another. But no

Farmers like Dan Jenniges see livestock grazing as a way to provide an economic incentive for keeping more perennial cover on the landscape.

matter what Jenniges is inhaling, it's clear his willingness to push the envelope has kept him on the land at a time when more "conventional" operators have failed. Sometimes his endeavors aren't directly connected to agriculture—during a rough patch his farm business was going through in the 1980s, Jenniges kept himself from taking a town job by hunting coons and trapping leeches to sell as bait. "It was surviving is what it was," he said.

Jenniges is also willing to innovate within an innovation. At one point during my tour, we pulled into a pasture, making the first of what turned out to be many almost impossible traverses through steep, rut-filled terrain with a questionable ability to keep his 1996 GMC truck afloat—it had just rained heavy the day before and after a stretch of muggy weather a cool front was moving in, gently bending the pasture grasses. "I guess it did rain," Jenniges said nonchalantly as we started to bog down in one low spot and the engine struggled mightily to power us out.

We stopped at a high spot and got out to take a look at a piece of ground that represented what can happen when a farmer and the environmental community join forces to support a shared resource. Here on a hilly stretch of pasture abutting the East Branch of the Chippewa River, Jenniges had just launched an initiative that blended cutting-edge grazing and soil improvement management with ecological restoration techniques. His goal was not only to increase production during the prime part of the growing season when cool season grasses do well, but to extend the grazing season through the hot weeks in late summer when such species often go dormant. On part of the pasture he had planted warm season native grasses and forbs, such as sideoats grama, Indiangrass, big bluestem, little bluestem, and white and purple prairie clovers. To prepare the seedbed for the native species, he had planted a cocktail mix of cover crops that included, among other species, soybeans, turnips, sorghum, oats, and field peas, which he was also able to graze before he seeded the prairie species. This diversity of annuals helped build the fertility and general biological activity present in the soil. Soil scientists have described such a method of preparing a seedbed as "jump-starting" or "awakening" the soil biome.

Jenniges had also expanded the number of paddocks in the pasture from four to eighteen. This allowed him to rotate the cattle more frequently, giving the forages in each paddock more time to recover—as much as sixty to eighty days—and distributing manure and urine evenly throughout the landscape. Moving cattle through paddocks on a regular basis is a key element of managed rotational grazing, but Jenniges had taken it one step further. He is a recent convert to "mob grazing," which relies on putting a relatively large number of cattle in a paddock for a short amount of time—sometimes less than a day. Since the animals are moved so frequently, they often end up leaving behind 50 percent or more of the available forage. The animals end up trampling a lot of the vegetation into the ground—it's akin to getting up from the supper table with leftovers sitting on the plate, and then stepping on them for good measure so you won't be tempted to eat later. For a livestock producer used to making use of every last bit of nutrition available on the land, such a strategy can seem wasteful, even lazy. But the method is based on the idea that stamping that biomass into the ground helps feed a different herd—the one residing underground. It's been said that if one was to place on a scale all the bacteria in an acre of topsoil, it would weigh at least as much as a cow grazing on the surface.[12] All those organisms feed on not only the cattle's manure and urine but the vegetative matter that's hoofed in. This regenerates the pasture even faster and provides long-term resilience.

Jenniges admits this mob grazing system was difficult to accept at first, given his propensity to harvest as much grass as possible off a pasture. In fact, he's compared it to being as much of a challenge as giving up smoking. "I'm used to the idea of, if it's there, you take it," he said. "It's really a mindset to realize that if I leave this I'm going to gain something back from it."

When I first visited this revamped pasture one summer evening, all these changes to Jenniges's grazing system were toeing that line between "innovative" and "will it work?" The farmer was still getting his cattle watering system established (no small investment—it required the drilling of a new well) and his cocktail mix of cover crops was still in the process of preparing the seed bed for the planting of perennial grasses and forbs. But when I returned to this spot with a group of farmers and natural resource professionals a year later, it was clear the gamble had paid off. "This is the most grass I've ever seen on this pasture," remarked Jeff Duchene, a local grazing expert for the U.S. Department of Agriculture's Natural Resources Conservation Service.

The farmer had actually been able to increase the land's carrying capacity significantly by bumping up the number of brood cows grazing on this 189-acre piece of land from sixty-five to eighty-five. Jenniges, like other livestock producers in the area, is extremely concerned about having access to enough land to support a profitable herd size. Producing more with less real estate was one way to get around the problem of needing a certain number of cattle to make a go of it financially. And yet, he wasn't willing to push the land beyond its limits because of his own financial goals. I asked Jenniges how much he hoped to increase beef production with the smaller paddocks.

"I guess the grass will tell me that," he said. "It'd be great to run 180 cattle up here, but I'm realistic enough to know that I can't do it all at once. I'll have to do it over a number of years."

I believed Jenniges when he said that he would pay attention to the capacity of his grass to support more hooves. He's a first-rate observer. The first time I met him was at a public tour of a prairie owned by The Nature Conservancy. Called a "BioBlitz," the event was a kind of nature-based scavenger hunt that allowed community members to tabulate the number of plants, birds, mammals, amphibians, and insects present in the prairie over a twenty-four-hour period. During the event, Jenniges was naming so many plants that the naturalist leading one of the nature walks began deferring to him when a stumper emerged.

"I should know one or two, shouldn't I?" the farmer said curtly when I later complimented him on his knowledge of native grasses and forbs that I had noticed at the BioBlitz. In other words, a guy should know the resource he relies on for a living. While he said this, he pointed out that many of the same species that are present in The Nature Conservancy's prairie grow on his own land. "A lot of the plants that people on the BioBlitz thought were unique or nice to see, well, a lot of those plants are here."

Dan Jenniges used cover crops to prepare soil for the establishment of a perennial pasture mix that included native grasses and forbs.

Even more important, he knew the relationships between plants, the land, and animals. At one point during the BioBlitz, Jenniges explained to others on the plant identification hike how cattle graze sideoats grama at the base and leave the stem, which is the seed producer, alone—a nice collaborative relationship that preserves a warm season grass needed for grazing during the hottest part of summer. By the time the sideoats grama is thriving, cool season grasses like brome have already seeded out and are dormant, he explained before walking away from the group of nature lovers to light up a cigarette. Later, a prairie plant expert identified some echinacea, otherwise known as purple coneflower, and asked Jenniges, "Will graziers tolerate echinacea?"

"Tolerate some of everything," Jenniges responded with a shrug.

Back at Jenniges's revamped pasture, the Fish and Wildlife Service's J. B. Bright made it clear the livestock producer was doing more than tolerating a few native species here and there. Like Duchene, he liked what he saw, but for different reasons. He noted how the rotational grazing system had created structural diversity, or heterogeneity, over the 189 acres. In other words, vegetation was at different heights and levels of maturity across the landscape. This is critical for providing different species of wildlife the kinds of habitat and food sources each requires throughout the growing season. The needs of grassland songbirds differ from that of mallards or Canada geese. Bumblebees seek out different flowers than other pollinators.

It's that kind of diversity Bright strives for on the refuge acres he manages, including a waterfowl production area that borders the Jenniges pasture. That's why, working with livestock producers like Jenniges, he utilizes grazing to manage grasslands on refuge lands. And there's an added benefit to this relationship: better community relations. Jenniges explained that growing up in an area where there are a relatively high number of state and federal wildlife areas generated a negative view of refuge lands among livestock producers. While scrambling to find enough pasture to graze on, they watched helplessly as these public areas become overgrown with invasives. Giving livestock producers periodic access to these lands shows that natural resource professionals such as Bright acknowledge the role working lands conservation can play throughout the community. "It gives me a reason to accept some of these public lands in our community," said Jenniges.

Such improved relations don't just make for a less tense community. Jenniges's planting of warm season natives in his pasture was done with the assistance of habitat specialists like Bright. For good reason: those species not only help get cattle through a summer slump when cool season grasses tend to go dormant, they also provide food and habitat throughout the growing season for wild residents. There had been a bit of cross-seeding between the refuge and the farm, as well as between the refuge manager and the farmer, and on this day both sides acknowledged a benefit. Bright and Jenniges obviously have a relationship that is not always completely amicable, but is built on mutual respect and the sharing of some mutual goals.

"The first time I met J. B. was when he stopped and asked to hunt my place," Jenniges told me one day after a grazing workshop on a neighboring farm. Bright was seated on a pickup tailgate next to the farmer as he said this.

"No! I wanted to know how to get to the wildlife production area!" Bright blurted out good-naturedly.

"Somehow or other he asked about hunting," Jenniges maintained, as he lit up another Doral. "I told him to hunt on his own damn land."

Bright put a more diplomatic spin on that initial conversation. "What he said was, 'I don't think you're doing a very good job with your grass management,' and I said, 'I 100 percent agree.' At that point, I had walked on enough land of ours that I realized, we've got to step it up."

Thus was born a relationship based on an acknowledgement that buried behind some past animosities, or, as Jenniges calls them, "feuds," between local farmers and government natural resources agencies, was a common goal: keeping the land in continuous living cover.

Such relationships don't survive and thrive by accident. Both Bright and Jenniges are participants in an initiative called the Chippewa 10% Project, which is working in the Chippewa River watershed to find ways for farmers and other landowners to get more continuous living cover established in ways that are profitable. Coordinated by the Land Stewardship Project and the Chippewa River Watershed Project, the initiative was sparked by research showing that even if just 10 percent more of the watershed was converted from corn and soybeans to perennial systems like grasses and forbs, water quality would improve to the point where it would pass basic state environmental standards.[13] Through meetings and one-to-one conversations facilitated by Chippewa 10% staffers, it was determined that farmers and environmentalists have a common desire to maintain more grass in the area. Some of these discussions also involved what were called "recreational landowners" — people who had bought real estate in the area with the thinking that if they would just leave it idle, it would become prime hunting habitat. Instead, many of these plots, lacking disturbance, simply became overgrown with sumac, red cedar, and buckthorn. Because of the Chippewa 10% Project, some of these landowners are now seeking out livestock graziers to help provide the disturbance they needed to bring back mixed grassland habitat.

Graziers Without Borders

One of the things Jenniges mulls over when checking his pastures and building fence is an idea that could further bring together as a kind of community partnership all those individuals and groups who want more grass on the landscape. The ultimate goal, which was formulated by farmers, The Nature Conservancy, and other partners in the Chippewa 10% Project, is to combine many smaller cattle herds that could then be moved across public and private property lines in long-term rotations, providing the right mix of large-scale impact and rest natural habitat requires, while giving livestock producers flexibility and access to perennial forage.

It's believed such an initiative would not only expand the benefits of conservation grazing beyond refuge boundaries but would make private, nonfarming landowners a part of this team effort to save habitat. Steve Chaplin, The Nature Conservancy scientist, calls such a concept "coordinated landscape management"—it's a way to prevent the creation of islands of habitat that are overwhelmed by bad land use throughout the rest of the region. I prefer the simpler term: community conservation.

"By having a mixture of private and public lands managed well, we can have a wider landscape level impact," Chaplin said. "We need to talk about the overall landscape and not just a particular plot of ground."

With all this talk about how fences serve as the basis for a grass-based landscape, this initiative would involve knocking down such barriers, at least metaphorically. Jenniges, always thinking big, sees an opportunity where farmers and nonfarmers could be a part of a common marketing cooperative in which each member owns a percentage of the livestock being used to manage the landscape. Such a cooperative would not only help bring together the large numbers of animals needed to manage a big expanse of land, but could provide natural, grass-fed meat and other products to consumers who want to know their food choices support healthy habitat. Through such an effort, another group of people could be drafted into a community effort to create more resiliency: conscientious eaters. This could have a trickle-down effect. More cattle being marketed directly, for example, means a local locker plant stays busy processing meat, creating economic activity year-round.

"That kind of activity starts to add up," said Jenniges. "Somebody coming hunting for a few months in the fall isn't going to do it. It's not going to support schools, churches, and businesses the rest of the year."

Interwoven with Jenniges's independent streak are strands of a deep, abiding love for the farms, towns, and people that make up his part of the world. By ignoring all types of boundaries, he's found a way to connect the natural elements of community with the wilder ones. One summer evening, the grazier took me to the tree farm in his neighborhood where he got his start in livestock production. Grasses, including red top, big bluestem, little bluestem, and sideoats grama, covered a steep hillside that almost blanked out the sky. Draws off to the side were covered with evergreens. The land had been put into a perpetual grassland easement by the owners, which means corn, soybeans, or any other annual crop are not part of its future. As Dan's GMC ground up the steep hill, he said, "You ever get to the point where you're just, I don't know, frustrated and just need to go sit some place for a while? I'll show you my spot." We crested the hill, and a breathtaking view of pasture fell away to a line of tall trees along a lake. A herd of first-calf heifers sauntered up the slope to greet Dan as he walked among them. "It's just a place where you don't see no corn," he said with that lopsided grin before turning back to the cattle. "Do your work, girls."

4

Brotherhood of the Bobolink

In Search of the Ultimate Indicator Species

Avoid looking across the road to the west, where lies an apocalyptic vision of what is and what could have been. Rather, look to the east, where there's ecological potential and a reminder of why it's worth forging links between birds, bovines, and biologically healthy soil. That's the version of a mantra Mary Damm recites to herself each time she pulls into the driveway of the farm she owns in the Driftless Area of northeastern Iowa, bleary-eyed from a 475-mile road trip that begins at her home in Indiana.

Damm purchased these 120 acres in 2014. As I describe in the introduction to this book, the previous owner, Dan Specht, had been killed in a haying accident the summer before. Before the accident, Mary had spent much of the previous decade wandering this mix of pasture, trees, and restored prairie, helping Dan tabulate all the birds and plants the farmer had nurtured on this wildly successful operation while raising crops and livestock. So when the property, which Dan dubbed Prairie Quest Farm, came up for sale after his death, she couldn't bear to see it bought up and plowed down for corn and soybeans. Damm's fears were warranted: she had lost a bid to purchase an additional twenty acres of Dan's land immediately west and across the road from the main farmstead. The first time she visited her farm after the auction, the acrid odor of smoldering slash piles hung in the air—the new owner of the twenty acres had bulldozed a quarter-mile line of trees and piled them up for burning; the pasture that had bordered the trees was being prepared for row crops.

"I cried, and I told Dan I was really sorry that I couldn't buy that land," Damm recalled on an afternoon in late July as we stood in her driveway, looking at those twenty acres, which were now head-high corn from fenceline to fenceline. Then she turned east to look at the 120 acres she had saved from the dozer and the plow. The grasslands that made up most of Prairie Quest Farm were speckled here and there with small flags and crisscrossed with portable electric fencing—the former marked research plots, the latter rotationally grazed pasture paddocks. The flags and the fencing represented a possible way to not only maintain the legacy of Dan's farm as a home to healthy ecosystems, but give other farmers and the rest of society a reason for not always seeing a stand of trees or grass as unproductive until it's dozed, burned, and plowed.

In recent years, I've visited Prairie Quest a couple of times to learn more about a research collaboration that has arisen in the wake of Dan's tragic death. Mary, a prairie ecologist, is partnering with Phil Specht, Dan's older brother, who owns and operates a dairy farm near her land. As I toured Mary and Phil's farms, the whole time listening to them compare notes, debate, and downright argue about the way to strike a balance between scientific veracity, environmental sustainability, and agricultural profitability, I realized that on a micro-scale they are grappling with a question that vexes our entire food and farming system: how do we develop an indicator, a kind of label, that immediately relates a clear message about the impact a farming method is having on ecosystem health and at the same time gives the public a helpful clue as to what it can do to support that type of agriculture?

I could think of no better piece of land to contemplate that question on. After all, no ecological agrarian was more aware than Dan Specht of the push and pull required to strike the working lands conservation balance.

Prairie Partners

This collaboration has its roots in a chance meeting Mary had with Dan at the 2004 North American Prairie Conference in Madison, Wisconsin. Although the tallgrass prairie biome was once present in fourteen states from Texas to Minnesota—including 85 percent of Iowa—sprawl, agriculture, and other forms of development have combined to all but obliterate this habitat in the midwestern Corn Belt. Iowa, Illinois, and Wisconsin, for example, now have less than 0.1 percent of their presettlement prairie.[1] As a result, prairie ecologists are particularly interested in two types of prairies that exist today: native remnants that somehow escaped the plow and the bulldozer, and restored prairies—plots that have been replanted by land trusts, natural resource agencies, and private individuals using seed gleaned from those remnants and native plant nurseries.

As part of her graduate school research at Indiana University, Damm was interested in comparing the ecosystem health of the native remnants and their restored counterparts. After the meeting in Madison, Dan invited her to study prairies in Iowa. They became romantically involved and soon Mary was regularly making trips from Indiana to gather samples for her research. By 2017, she was closing in on finishing her PhD dissertation, which was showing that although restored prairies are not as diverse as their native counterparts, over time the groupings of the restored habitat's plant communities in a given location become more consistent from year to year, which makes the overall system more resilient. Resiliency is the hallmark of a working, or functioning, ecosystem like a prairie.

But it wasn't just about prairies that Damm was learning when she made those early research trips to Iowa. While using Dan's farm as a home base, she had her eyes opened to how agriculture could relate to nature in a positive way.

"I had been to two farms before I met Dan. I actually did not know much about midwestern agriculture," recalled Damm, who, before attending graduate school, worked for The Nature Conservancy and the National Park Service in Colorado.

Even when she first started visiting Dan's farm and saw how, for example, he had replaced some of his annual row-cropped acres with perennial pastures so he could produce beef cattle on grass, Mary was more drawn to the acres of oak and other hardwoods that bordered the pastures. "We'd hike in the woods and I would separate the woods from the pastures. I'd be like, 'Oh my, look at the beautiful forest understory,'" recalled Damm.

Eventually, it was the birds that convinced her working farmland and natural habitat could share the same piece of real estate. There were birds and other kinds of wildlife in those woods, of course, but

Mary also noticed that Dan's rotationally grazed pastures were home to an array of avian species usually associated with prairie habitats: various kinds of sparrows, as well as meadowlarks, dickcissels, and, perhaps the most noticeable of them all, the "skunk bird," otherwise known as the bobolink. Dan had been into birds since he was a kid, even making up comic strips about them and other wild animals. He always seemed to have a pair of binoculars on hand, and bought Mary her first pair. Chore time on Dan's farm was often not a straightforward affair.

"We'd be driving along and, 'Oh, oh, there's a bird!' And then we'd drive along some more and, 'Oh, oh!' So eventually we'd get back to what we were supposed to be doing, which was check Dan's cows," recalled Damm with a laugh. "The birds, we'd always look at the birds."

When Damm was back in Indiana, Dan would e-mail her about the seasonal arrivals of various species, and share his ideas for helping out the feathered residents of the farm, like adjusting grazing schedules or haying a field in such a way that the fledglings would have an opportunity to seek refuge from the mower. Mary and Dan soon became a kind of team—attending prairie and sustainable agriculture conferences, they would participate in different sessions, comparing notes afterward. Mary would accompany Dan on his cow chore/birding outings in the field and Dan, in turn, would help Mary do soil and plant sampling on her prairie plots; he even developed a height-adjustable plant sampling frame for her using electric fence posts.

Bobolink Battle

Dan bought Phil Specht his first pair of binoculars as well. Like his younger brother, Phil inherited their family's passion for the out-of-doors. Their dad had long been involved in soil conservation efforts on and off the farm they grew up on, and was an avid fisherman whose idea of a Sunday family outing was to be casting lines for smallmouth bass at a local river. Phil, who is in his late sixties, has a degree in social work with a minor in chemistry from Wartburg College. Dan studied wildlife biology at Iowa State before he decided, as Mary puts it, "I can go be a wildlife biologist at home." (In the 1990s, he went back to school and got his biology degree at the University of Northern Iowa.)

Both brothers ended up farming, and both decided they were not going to raise food using a conventional chemical- and energy-intensive system. In the 1970s, Phil started producing milk using managed rotational grazing, which at that time few in the Midwest had heard of. "I wanted my whole farm to be a working ecosystem," he told me.

Today, Specht has 250 acres of rotationally grazed pastures, and although he at one time grew as much as one thousand acres of corn, he hasn't raised a significant amount of that crop since 1992. Like Mary's farm, Phil's is extremely hilly, and years ago he put in place some seven miles of terraces in order to grow row crops on the highly erosive slopes. Today those structures, which snake along the contours of the farm, are covered in grass and trees.

Rotational grazing has proven to be an economically viable way to produce milk for Phil, but his interest in the land goes beyond the financial bottom line. One day when Mary and I met with him, he pulled up in front of Dan's now-abandoned house in an especially good mood because he had just sold one hundred woodland acres to the Iowa Department of Natural Resources. It was prime cerulean warbler habitat; these sky-blue songbirds rely on mature deciduous forests; one estimate is that their populations declined 74 percent between 1966 and 2015.[2] Phil was delighted he could help out a wild resident that he was fascinated with. "I got a *real* good look at one cerulean warbler," he told us. "It flew down, lit on a branch. I was staring at it, and then it gave me this beautiful look and off it went and I've never seen another one since."

He then handed Mary a couple of cover crop seed catalogs he had picked up. There was a one-acre spot near Dan's house that she had enrolled in a U.S. Department of Agriculture program that paid landowners to establish pollinator habitat. Damm wanted to seed it down with cover crops to prepare the soil before putting in the expensive prairie seed, and she admitted she was clueless about how to proceed; she could identify every plant present in an established prairie, but preparing the soil for a new planting was another thing altogether. Damm needed advice from the farmer.

They talked over the best strategy for establishing the prairie and the history of the farm, as well as government agriculture programs and some recent watershed-wide efforts to improve water quality and habitat. They also discussed Phil's approach to farming. "It should mimic nature," he said. That statement prompted Mary to share her philosophy on how she can best contribute to protecting and supporting the natural environment. "I became an ecologist because I wanted to protect these natural areas and my personality isn't that of a tree hugger who goes to meetings and expresses myself that way," she conceded. "So, I felt if I had the science knowledge, that would be my way to contribute."

Mary Damm and dairy farmer Phil Specht check out research plots in one of Damm's rotationally grazed pastures.

At one point during the discussion, the farmer grabbed a camera out of his van and tried to snap photos of an uncooperative butterfly that was flitting around a bull thistle a few yards away. "A female black morphed swallowtail," Phil announced. He takes a lot of photos of the birds, plants, and insects he sees on his farm and posts them to his Facebook page, along with his strident views about the state of industrial farming and agricultural policy. Specht was hoping his telephoto lens would help him differentiate between two male bobolinks he had been watching.

"Did you name them, Phil?" Mary teased.

"Well, I could've," Phil shot back without missing a beat.

Over the years, Dan and Phil partnered on various agricultural enterprises and traded work. All along, they shared an intense passion for farming in a way that was good for the land. Dan also adopted managed rotational grazing, in this case to produce beef cattle, and for a time raised pork utilizing a method reliant on deep straw bedding in open pens, a stark contrast to the factory-style confinement system that dominates hog production these days. He raised corn and organic soybeans, and when I first met him in the 1990s, he had just won a state yield contest for the latter crop.

No matter what farming technique they were using, the brothers were always on the lookout for indicators that their production methods were in sync with nature. So they were excited when, in the 1980s, Dan started noticing that grassland birds were nesting and feeding in the rotationally grazed pastures. "All of sudden, there are *lots* of birds," recalled Phil.

Others have noticed as well. While doing cerulean warbler research along nearby Bloody Run Creek a few years ago, natural resource scientist Paul Skrade would regularly take shortcuts through Phil's farm. He was, in his words, "blown away" by all the grassland birds he was seeing on the farm. Sighting so many bobolinks on Phil's land seemed to be particularly delightful for the assistant professor of biology at Upper Iowa University, who describes one of their calls as resembling the musical beeps emitted by the robot R2-D2 in the *Star Wars* movies. Bobolinks and other birds reliant on upland grassland habitat have experienced the biggest decline of any bird group in North America, and the downward trend shows no sign of dissipating. Replacing all those pastures, hayfields, and grassy field borders with annual row crops has had a particularly detrimental impact on such birds.[3]

"I was seeing all these bird species that are of concern in Iowa while walking across Phil's farm," the biologist recalled. "I said, 'Phil, what's the deal? Is this Conservation Reserve Program ground?'"

Phil made it clear that this was not idled conservation land—it was a working pasture system.

Birds such as bobolinks are "obligate species"—meaning they rely almost 100 percent on a certain kind of habitat—grasslands, in this case. Finding such habitat when they return to the Midwest each spring is critical to the skunk bird's survival. They winter in South America and after making a jaw-dropping, 12,500-mile round-trip migration flight each year, begin building their nests in places like northeastern Iowa by the second or third week of May.

One afternoon while he was visiting Mary's farm as part of a Practical Farmers of Iowa field day, Skrade explained to me that just having lots of grass isn't enough—grassland songbirds rely on a heterogeneous habitat; they want variety both in terms of the height of vegetation *and* the number of plant species present. Skrade likes that Phil does not have a uniform way of grazing his paddocks. Depending on conditions and time of year, sometimes the farmer leaves the cattle on the same spot for a couple days and they are allowed to eat the forage down relatively short. Other times the cattle may be moved after only a day. This latter

Prairie ecologist Mary Damm and ornithologist Paul Skrade discuss how grazing is impacting her farm's grassland habitat.

method leaves a fair bit of standing vegetation behind, allowing the paddock to recover quicker *and* providing a diverse habitat for birds and other wildlife.

"Agriculture and biodiversity can go together—we're seeing that here. As you get this diverse habitat out there, you can get this diversity of species," said Skrade excitedly as a couple of dickcissels somewhere out in a nearby pasture cranked out their *dick, dick, dick . . . cissel* call incessantly. "We have a working landscape here."

I realized while talking to Skrade that he was referring not only to the fact that Specht's pastures were generating economic activity, but that they were also part of a system that "worked" from an ecological point of view. Such an insight recasts how one thinks about a "working landscape"—it's not just about pounds of milk per acre; ecological services are a valuable output as well.

When Dan got Phil those binoculars, it launched a kind of avian contest between the two that continued even after Dan's death. Both men were fascinated by the charismatic bobolink, so at the core of the competition was a basic question: whose farm could produce the most of that species in a given year? During one of my visits to his farm, Phil estimated that of the 250 acres of pasture on his place, about one hundred acres were prime bobolink habitat; there were some thirty nesting pairs of bobolinks on his farm alone.

This wasn't an assessment based on off-the-cuff observations made from a tractor seat. For the past few years, Specht has been in the midst of a fairly in-depth research project of his own. Phil is not your typical farmer—besides having academic training in chemistry, he's a member of the American Association for the Advancement of Science, which publishes the journal *Science*. One of his "hobbies" is reading scientific abstracts and he complains about having to pay money every time he wants to download journal research that was conducted by land grant scientists.

"He's a weirdo," said Mary with a deep-throated laugh.

Specht is a big believer in recording data and adapting his management style as a result. He makes sure the information gathered is randomized, "because your eyes will always go to the exception," he said. Such an observe-and-adjust way of doing things is relatively common among farmers I've met who are undertaking innovative practices that replace a heavy reliance on chemical inputs and machinery with intensive management. Over the years, I've been on dozens of operations that utilize managed rotational grazing, and a trait graziers share is the ability to monitor how their livestock and grasslands are interacting, and to act accordingly. Managed rotational grazing, like diverse crop rotations or cover cropping, is not a cookie-cutter way of managing the land that lends itself to computerized calibrations. Phil likes to quote a grandfather of his who used to say, "The grass will wave and beckon the cattle." In other words, when it's tall enough that the breeze can cause movement of the stems, it's tall enough to be grazed. "My number one rule for graziers: observe and adapt, observe and adapt."

Phil has set up rectangular study plots on his and Mary's pastures. For two months each spring, the farmer notes when female bobolinks are flushed off nests in a plot. He also records grass height at various times during the study period using a measuring tape attached to a five-gallon bucket. "My thinking is you're not going to get farmers to think about fine details, but you could get them to set a bucket out to measure grass height," said Specht. "That's why I went with the bucket—everybody's got a bucket."

Bobolinks forage for insects and spiders found around forbs, grasses, and sedges. What Specht has observed is that the birds prefer to nest in paddocks that have been grazed in early May, and that they stay away from the parts of the pasture that border the woodlands. The farmer's ultimate goal is to figure out at what point livestock productivity and bobolink productivity begin to intersect, or collide, depending on how you look at it. How much forage production can he get off his pastures before the bird nesting suffers? What is the tipping point?

"What I think I'm proving is you can have a healthy grass ecosystem while producing milk," said Phil. "But I'm also measuring why are the birds in one spot, and not another? What do they prefer?"

Damm's research on her and Phil's farm is going deeper, so to speak. Along one-hundred-foot transects in the grazed pastures and the ten acres of prairie Dan restored on a back part of the farm in the late 2000s, she records how many plant species are present and each species' abundance, thus developing a picture of how much diversity there is. She also takes regular soil samples, sending them off to a laboratory to be tested using a sophisticated method called the Haney Soil Health Test.[4] Such testing goes beyond the traditional measurement of basic nutrients like nitrogen, phosphorus, and potassium—known as N-P-K—which is popular among crop farmers trying to determine their fertilizer budget for the coming year. Comprehensive soil health tests attempt to show how much biological activity is actually going on. Such a measurement provides insights into more than a soil's fertility; it also measures its ability to, among other things, build soil organic matter, which in turn can bolster everything from water-holding capacity to the amount of greenhouse gases that are sequestered. Damm is excited about making connections between the health of the soil and the state of the plant community that is sending its roots through it. After all, one truism that's come out of the soil health movement is that diversity below ground equals diversity above ground.[5]

The study Damm is undertaking on her farm is similar to the dissertation research she's been doing on restored and native prairies in other parts of the state. In other words, will that prairie Dan planted become more resilient, more prairie-like, with time? But the question has an added twist: what impact will the grazing of the pastures and parts of the prairie have on species diversity and soil health?

Damm foresees the possibility of a joint scientific paper with Phil that compares ecosystem health between a grazed pasture and a restored prairie, and that takes into account the impact rotational grazing has on

plant diversity, soil health, and bobolink populations. Such a collaboration makes sense in a lot of ways. Because Phil's farm is practically next
door to Mary's (another landowner's fields separate the two), their individual patches of grasslands and woods form an almost contiguous block
of natural habitat.

Pasture vs. Prairie

One summer day we three headed up to Mary's pastures to get a firsthand look. There were sixteen rotational paddocks on one hundred
acres, and we stood in a spot that had just been grazed a month previous by a local grass-based beef farmer to whom Mary was renting the
pasture. From my non-expert perspective, it appeared to be a well-
managed grassland: good regrowth, no exposed soil. Flags marked where
Mary had been sampling vegetation and soil, as well as where Phil had
been doing bobolink monitoring. We spied a few female bobolinks, which
were buff-colored, a sharp contrast to the breeding males, which, with
their feather color mix of black below and white above, suggest "a dress
suit on backward," writes Roger Tory Peterson in the first birding field
guide I ever owned.[6] We spied one bobolink Phil believed was a late
nesting female, judging how much she held her ground and energeti-
cally let the neighborhood know we were in the vicinity with her *chuk,
chuk* call.

I asked Phil and Mary to assess the pasture from their various stand-
points. Phil said the pasture had some good forage in it from a livestock
producer's perspective: timothy grass, orchard grass, bromegrass, red
clover, and white clover. There were also stands of giant ragweed and
goldenrod, which are not such good sources of livestock feed.

Mary conceded she didn't have the eye for pasture that she does for
prairie.

"Look! Goldenrod! Look! Milkweed!" Phil shouted, doing an im-
personation of what he thought a prairie ecologist's response to the
habitat would be, a reminder that even though a plant like goldenrod is
a valuable part of the natural habitat—it's a great source of food for
pollinators during the fall, for example—it's not well-loved by cattle.

"Actually yes, I do see the diversity of plants. Just on the other side of
the fence where I sampled one bird plot, there were thirteen species of
plants," Mary said, pointing at a spot a few hundred feet away. "There
is a diversity—it's not as diverse as a native prairie or maybe even a

restored prairie, but in terms of what plants are out here and how many there are, and the cover too, it's good. So, one other thing I notice here— I know this is a cool-season grassland that this time of year has dead seedheads. And then over here straight up ahead of us is the prairie."

As she said this we glanced at the ten-acre restored prairie a few hundred yards east of where we were standing. The difference was striking: its warm-season natives were green and vibrant, and various flowering plants were just coming into their own in the July heat. Here in late summer were two habitats going in opposite directions.

Looking for Signs

"People like Dan and Phil don't get caught up in the minutiae, or they see the minutiae and go beyond it and integrate all the minutiaes," Damm said later that day after Phil had headed back to do the evening milking. "I think unfortunately scientists get so caught up in a little piece of the picture that they are not very good at anything big."

As she said this, we were walking down to the ten acres of restored prairie. It was hemmed in on three sides by trees. As we made our way down the path, more bobolinks flashed about, giving out their soft *chuk* call when perched, and emitting the R2-D2 song in flight.

We waded waist deep into the prairie, and it was clear Mary was more in her element. All of the talk about cover crop mixes, government programs, pounds-of-bovines-per-acre, and fencing systems was part of a steep learning curve for her. She relied heavily on Phil and other farmers in the area for advice on what to do with the agricultural part of the land. But here, among the native grasses and forbs, she was in charge. Along the way, Damm pointed out where she was sampling plants and soil and how she had unofficially divided the natural habitat into "bad prairie" and "good prairie." The former, which was on the side of a small pond closest to the pasture, was full of brome, a good forage for cattle but which is considered an invasive by prairie enthusiasts.

At one point, I asked to take a photo of her in the prairie, and Mary agreed, as long as bromegrass wasn't in the shot. "Phil loves brome," she conceded. But the farmer wasn't here and the prairie ecologist was. We struggled through reed canary grass as a bullfrog near the pond launched into its evening amphibian public address system. Finally, we made it to the "good prairie." There wasn't nearly as much bromegrass, and there was a healthy mix of wild species such as big bluestem,

Indiangrass, Canada and rigid goldenrod, gray-headed coneflower, cup plant, rattlesnake master, white wild indigo, round-headed bush clover, showy tick-trefoil, partridge pea, and common mountain mint. Damm stopped to examine a legume that was doing particularly well: the vines of *Apios americana*, commonly known as potato bean or American groundnut, were wrapping themselves around the stems of reed canary grass. A small native legume was quietly taking down a highly productive, invasive grass.

A Tough Neighborhood

Mary is well aware that no matter what she does on this farm, or what Phil does on his, they are just two of many in the region—islands in the stream, or, more accurately, islands in a roaring river that frequently leaves its banks. All she has to do is look west across the road at that twenty acres of former timber and pasture to be reminded of that. The bottom line is the land management practices that dominate the rest of the agricultural landscape know no boundaries.

For example, as part of her research Damm studied two-hundred-acre Steele Prairie in northwestern Iowa's Cherokee County, which was at one time home to the state's largest population of western prairie fringed orchid. Unfortunately, it is surrounded by some of the most expensive farmland in the nation. A few years ago, Damm and a botanist conducted an inventory of the prairie's flora and found *no* western prairie fringed orchids. Over the years, eroded soil from surrounding row-cropped fields had crept into the prairie, bringing with it the fertilizer present in those crop fields. Such an influx of sediment and nutrients favors invasives like reed canary grass, which crowd out native species like the orchid. It was a great loss. Damm compared the orchid to the bobolink—a charismatic species whose presence or absence tells us a lot about the health of an entire ecosystem.

But there are signs of hope. Several members of Practical Farmers of Iowa, a sustainable agriculture organization, farm in Damm and Specht's neighborhood, and frequently hold field days highlighting innovative, ecologically positive production practices. One such event on Mary's farm drew dozens of participants, despite torrential rains in the area that day. Phil knows of farmers in the area who are increasingly interested in grass-based livestock production, and conventional corn and soybean producers are more frequently showing up at sustainable agriculture

"My number one rule for graziers: observe and adapt, observe and adapt," says dairy farmer Phil Specht.

workshops, seeking information on cover cropping and other techniques to build soil health.

Damm also feels the environmental community is starting to appreciate the role working lands conservation can play in habitat restoration. Prior to one of my visits to the farm, she had an e-mail exchange with a leader in the ecological restoration community. Mary had suggested that the next meeting of the Midwest-Great Lakes Chapter of the Society for Ecological Restoration include a session on how to establish native prairie species in pastures. The ecologist had responded positively, writing: "I agree with you. I think it's time restoration and agriculture come more together."

Bringing those two worlds together is another reason the research she and Phil Specht are collaborating on may be so important. During one of my visits, they got into an energetic, but good-natured, argument over the best way to gauge whether a land management method like rotational grazing is facilitating a working ecosystem. Mary, the scientist, was excited about what something like a Haney Soil Health Test could show her. Certain Haney Test results could tell a lot about whether the farming system on the soil's surface was helping to build a diverse biome below.

Phil, the working farmer and observer, argued that just the presence of bobolinks can provide the telltale sign that things are working ecologically. It's an indicator farmers can note while doing chores, building fence, or riding the tractor; it doesn't require grubbing up a soil sample and sending it off to a lab.

"Do you have to do the Haney Test, or can you just see a bobolink? Dan's statement was, 'If you see bobolinks, three little words: It's. All. Working. Three words. You have a working ecosystem.' Anyway, the bobolink is just something you can take field glasses and see. It's a *real* handy indicator and a flashy one. Henslow's sparrows are more threatened than bobolinks, but frankly aren't as charismatic."

Damm agreed that the bobolink could be a recognizable symbol of how the land was faring, but as an indicator it needed to be backed up with other research that monitored more than successful fledges of just one bird species. That's why it was important to augment the nesting observations with the plant species sampling and soil biology tests. "We need to find out if there is some sort of correlation—there may or may not be," she said.

Ecosystem Seal of Approval

All this talk about developing an indicator of a healthy ecosystem brings up the larger issue of how to provide consistent economic incentives for wildly successful farming. Mary had been thinking more about that question as she considered the future of the land she had taken ownership of. There's a reason that the majority of farms in her community and beyond raise corn and soybeans: government programs and the markets pay them to do that. Although Damm was using U.S. Department of Agriculture conservation programs to provide financial support for improving the rotational grazing system, establishing pollinator habitat, conducting soil health sampling, and even establishing edible nut trees on her farm, such incentives are no substitute for long-term financial support. In addition, the survival of such programs is vulnerable to the whims of agricultural policymakers, and I've witnessed firsthand how the competency with which they are implemented through government agencies can vary considerably, depending on the staffing situations in rural offices.

As an absentee landowner with an environmental ethic, it's difficult for Mary to manage the 120 acres from afar, even with Phil being in the neighborhood. At one point, she had researched various templates for lease agreements that would require certain conservation practices be maintained by the farmer renting her land. In the end, Damm ended up enrolling the farm in a USDA Conservation Stewardship Program (CSP) contract. Such contracts, which usually run for five years, provide payments for maintaining various conservation practices on working farmland. Mary's CSP contract stipulates that the pastures be under a managed rotational grazing system and that it undergo monitoring of soil health, which dovetails nicely with the research she and Phil are doing. She has rented the pastures out to a farmer from the neighborhood who has an extensive background in environmentally friendly grazing, but that might not always be an option. Damm figures that for now, anyway, the CSP contract's stipulations will help keep in place environmentally sound land management practices no matter who the renter is.

Consistent market support for wildly successful farming practices is difficult to maintain as well. Certified organic farmers receive a premium price for maintaining certain chemical-free, sustainable practices—Dan was certified organic for a time—but establishing a prairie or building a healthy soil biota in general does not generate regular financial dividends in our current food and farm system. Over the years, Dan worked hard to try and get rewarded for his ecological farming methods through sustainable and organic meat labels, but had mixed results. Dan and Phil talked frequently about developing a sustainable grassland "stamp" featuring a singing songbird that could be placed on livestock products originating from farms that are doing the right thing when it comes to natural habitat. Maybe the bobolink could be the poster child for such a stamp, at least in the Midwest where so much of its habitat has suffered as a result of industrial farming? Phil wondered out loud about a third-party group certifying such a label.

This fixation on bobolinks as the ultimate symbol of a working ecosystem had me, like Mary, feeling a little uneasy at first. After all, any time we choose to focus on promoting one resource, there's the danger of excluding other pieces of the puzzle that are key to the workings of the whole. Aldo Leopold grappled with this when he realized that killing off predators like wolves may have been good for game species such as deer in the short term, but it resulted in significant damage to the overall forest ecosystem in the long view.[7]

However, as I spent more time with Damm and Specht, it became clear that the bobolink is so dependent on a healthy grassland
ecosystem—it relies on consistent, perennial cover that's characterized
by heterogeneity and is home to plenty of insect-based food—that using
it as a kind of biological barometer isn't a bad idea. With its black-and-
white flashiness, coupled with a name that tends to trip off the tongue,
the bobolink truly is charismatic and easy for even non-bird nerds to
remember. It may not be a keystone species, but its presence or absence
tells us a lot about what else finds that particular habitat attractive: other
grassland songbirds and pollinators, as well as the kind of deep-rooted
perennials that can keep our water clean and sequester greenhouse gases.
The bobolink is also one of those animals that can have a domino effect
when it comes to triggering one's interest in other members of the plant
and animal community. Partially because of his fixation on the bird, Phil
has become more interested in what insects are on the farm and ways he
can help them thrive. Mary, for her part, is now fascinated by the relationship between biologically healthy soil, plant diversity, and bobolink
nesting success.

Yes, we could do worse than to focus on what keeps this delightful
little bird happy. And Phil said something once that sticks with me,
providing hope that whether or not a special "Bobolink Beef" label is
created or enough credible science emerges from his and Mary's plots
and transects to gain the attention of government agencies and policymakers, one fact remains: the farmer and the ecologist are going to do
their utmost to maintain an ecosystem that Dan Specht would have recognized as healthy. Phil's reassuring statement came when I asked, half-
jokingly, who was winning the brotherly battle of the bobolink. "Me,
this year," he said without hesitation. "Last year, it was Dan."

The farmer then went on to describe ways he could tweak his grazing
system, tilting the odds even more in his favor.

5

Raising Expectations

A Team's Refusal to Accept
a Degraded Resource

When walking a stretch of North Dakota landscape under a withering summer sun, one's thoughts turn to moisture—or rather, the lack of it. So, when I and other participants in a farm tour kicked up indicators of cool, shady places while traipsing across a hayfield, it seemed like a mirage. Green-and-black leopard frogs were smoothly zigzagging out of our way, adding life to a field that had not gotten a decent rain in eight weeks. This part of south-central North Dakota was historically prairie pothole country, but no wetlands were in sight as wheat and corn stretched to the horizon.

"I've never seen so many frogs so far from a slough," said a fellow tour participant as we tripped through the field. "What's going on there that would bring them so far from cattails?"

When we reached the edge of the field where the couple who farm this land, Todd McPeak and Penny Meeker, were standing, they made it clear we weren't imagining things. "I hope you didn't step on any of my leopard frogs," Meeker said, smiling. We smiled too, and were especially relieved we hadn't injured any amphibians after she related a childhood story of using a stripped horse weed to "whip the crap" out of her brother and a cousin when she caught them shooting birds on their family's dairy farm.

Meeker and McPeak enjoy seeing birds, mammals, and frogs on the acres they produce grass, hay, cover crops, and beef cattle on. But these critters are also barometers of how the sustainable farming methods the couple use are affecting their business enterprise. As McPeak explained it, more frogs in a field connotes a healthier landscape that retains moisture in the soil more efficiently, which in turn translates into better-quality hay and grass that's drought tolerant. Such a healthy ecosystem is money in the bank when you're farming in a place that gets only around sixteen inches of precipitation a year. "From bees to badgers to beef, I see it all working together," he said.

While visiting McPeak and Meeker's livestock operation a few years ago, I was struck by a couple of things. For one, a wildly successful farm isn't just about thriving flora and fauna above ground—agroecology begins beneath our feet. In addition, when we raise our expectations of what a natural resource and the people who manage it can accomplish, the possibilities are wide open. Finally, true teamwork can be an incredibly powerful force for good, both on the land and in the communities that land supports.

In fact, it was because of an extraordinary team of farmers, conservationists, and scientists that I found myself in that frog-filled hayfield on a baking day. McPeak and Meeker belonged to the Burleigh County Soil Health Team, one of the best examples I've witnessed of how such a partnership can bring resiliency back to the land as well as the community. This group has spawned similar initiatives in Minnesota, Wisconsin, Iowa, and Indiana—as far away as Australia. My visit to the McPeak-Meeker operation was a follow-up—I had first spent time with the Soil Health Team the year before, and now I was back for more.

At the core of the Soil Health Team's work has been the promotion of practices that protect and regenerate the soil as much as possible. But it wasn't so much what innovative practices the team had successfully advanced that piqued my interest. Rather, what impressed me was how the team itself had managed to work together, providing a rich seedbed for testing, implementing, and supporting whatever creative sustainable farming practices were being tried. They had built a template for innovation and mutual support, which can be a critical tool for farmers who aren't exactly following the conventional path. New farming techniques come and go, but Burleigh County's Soil Health Team models the kind of environment needed to ensure that the roots for future innovations will always be deep and thriving. Every truly effective interdisciplinary team is made up of members who bring their own motives, skills, and worldviews to the table. To get at the heart of the Burleigh County Soil Health Team's success, let's look at it from the perspective of three members: a conservationist, a farmer, and a scientist.

The Frustrated Conservationist

At the core of this story is a change in attitude toward soil—perhaps one of the most taken-for-granted resources around. Consider, for example, how Jay Fuhrer used to do his job. For many years, Fuhrer was the Burleigh County district conservationist for the U.S. Department of Agriculture Natural Resources Conservation Service (NRCS). Burleigh County lies near the portion of the Missouri River that passes through the south-central part of North Dakota. Here the flatness of the state gives way to a more rolling landscape—a landscape known for wheat, hay, and "wild" pastures that contain native species such as big bluestem. During the 2000s, corn also become a bigger part of the farmscape here. Water is a dear resource in these parts, so for many years Fuhrer and other resource professionals focused on short-term efforts to get more water into the soil profile and keep it where plants could use it.

"We had accepted a degraded resource," Fuhrer recalled as he sat in his office in Bismarck on a September afternoon. "And when you accept a degraded resource you generally work from the viewpoint of minimizing the loss. And so we would apply a lot of practices."

Fuhrer's specialty during the 1980s and early 1990s was putting in grassed waterways in an attempt to keep water from running off so quickly. It helped, but didn't get at the core of the issue: why was that water not infiltrating the soil in the first place? What farmers and soil scientists in the area were starting to figure out was that the production system that had come to predominate—extensive tillage, low crop diversity, no cover crops, livestock kept out all season long on overgrazed pastures—was compacting the soil to the point where little water could make its way beneath the surface. It was also sharply reducing the amount of soil organic matter that was present. That's a big deal: organic matter is the energy-rich portion of the soil profile that's made up of plant and animal residue, along with the tissues of living and dead microorganisms. It controls everything from how much nutrition plants get to the amount of water that makes its way through the soil profile. Since organic matter is around 58 percent carbon, it also determines how much of that element is present in the soil; that makes organic matter a major player in sequestering greenhouse gases such as carbon dioxide. In short, organic matter drives the entire soil food web.[1] Unbroken prairie soils can have as much as 10 percent to 15 percent organic matter. But because of intensive tillage, midwestern soil organic matter levels have plummeted to below 2 percent of total soil volume in some cases.[2] This means the soil has little opportunity to cook up its own fertility via the exchange of nutrients, making it increasingly dependent on applications of petroleum-based fertilizers.[3]

To deal with the water infiltration issue, the Burleigh County Soil Conservation District's supervisors eventually formed a team that consisted of farmers and conservationists. From the beginning, the team promoted no-till, crop diversification, and simple cover crop mixtures. It also worked to get farmers to replace the traditional technique of turning cattle out into large pastures all season long with rotational grazing systems. These farming techniques have proven to be a vast improvement over intense tillage, monocropping, and overgrazing. Thanks in part to the Burleigh County Soil Conservation District's soil health work, at one point 70 percent of the county's farmers had adopted no-till cropping systems, which has to be one of the highest percentages in the country. But Fuhrer and others were finding that even with these conservation improvements, soil was still lost, precious water ran off increasingly compacted fields, and the quality of crops and grasses being grown kept deteriorating.

What was needed was a way to test out new approaches to building soil health while spreading that information among farmers as quickly and effectively as possible. One way the District has done this is through experiments at Menoken Farm, a 150-acre educational site started in 2009. Replicated trials on cropping practices that build soil health are done at Menoken and the District shares the results through field days, workshops, and a website.[4]

But Fuhrer and others know that farmers need to see these practices put into action on real working farms, ones that share the same soil type, geography, weather, and even economic conditions. So the District started promoting "25-acre grants" for seed. The farmers used the grants to establish cover crops and in return for receiving the free seed, the producers would serve as one of the stops on the Soil Health Tour, an annual end-of-summer event. Those twenty-five-acre test plots were popular, with the District overseeing thirty to forty a year from 2006 to 2008. With the price of cover crop seed at thirty to thirty-five dollars an acre, it was a bargain in terms of the harvest of real-world results it produced.

USDA soil health expert Jay Fuhrer shows how cover cropping and grazing can create good soil structure during a Burleigh County Soil Health Team field day.

"So part of the bargain was a willingness to speak at the tour stop— what worked, maybe what didn't work, their observations," said Fuhrer while going over test plot results in his office. "And then at the same time it gave people like myself the opportunity to take a look at those soils, maybe do an infiltration test on them. It allowed us to kind of ride along and monitor that and really kind of look at the benefits."

That created a whole lot of on-the-ground results with a relatively small financial risk on the part of the farmer. It also developed an environment where farmers were comfortable sharing their experiences— both good and bad.

A combination of results from the Menoken Farm and the fields planted using the twenty-five-acre grants showed that cover cropping— a system where non-cash crops such as cereal rye and tillage radish are grown in fields between the regular cash crop growing seasons—could build soil health year-round, not just during the spring and fall. The Soil Conservation District and the farmers also learned that diverse seed mixes that went beyond the traditional cover crop plantings of small grains and brassicas built up an impressive amount of carbon while feeding microbes. This makes soil naturally fertile and less reliant on chemical inputs, as well as increasingly erosion- and drought-proof. In other words, the soil is more resilient. And this resiliency can be attained relatively cheaply by seeding cover crops—plants that, by the way, can serve double duty as livestock forage.

One cover crop breakthrough Fuhrer and the other Soil Health Team members had was to ignore old ideas about plant competition. One year the soil conservationist was monitoring eight different species of cover crops planted on test plots. In one plot each species had been planted as a monoculture, and the other plots contained various combinations: two-way mix, three-way, etc., all the way up to where all eight species were planted together—something called a "cocktail mix."

"And then we had one of the driest years on record," recalled Fuhrer. "And I just thought, 'Oh, everything's failed and we're just not going to learn anything this year.' And I was so wrong."

What they learned was that the monocultures failed, and the mixes involving just a few species didn't fare much better. But the eight-way cocktail mixture didn't seem drought-stressed at all, and in fact yielded quite well. "It really taught us a lot from the viewpoint of how plants won't necessarily compete with each other—they can actually help each other," he said.

The idea that diversity is strength takes a page out of nature's notebook. Ecologists have found that in planted prairies, greater diversity results in a similar synergistic effect—making the entire system more resilient.[5] Fuhrer and his colleagues started thinking that the key to creating more resilient soil was to build more organic carbon down below. It's been said that soil without biology is just geology—an accumulation of lifeless materials unable to spawn healthy plant growth. That biology is driven by organic carbon, and socking it away for a rainy day (or a very dry one) pays big dividends. Those eight species of plants growing above ground may appear to have the potential to be in competition, but all the while their roots are interacting and serving as the basis for an incredibly diverse subterranean ecosystem.

Fuhrer and other soil conservation experts in the region were particularly impressed with results some farmers were getting by combining cover cropping, livestock impact, and no-till agriculture in a way that soil health could actually be improved, not just maintained at a high enough level to grow a stand of wheat or corn. For Fuhrer, taking such proactive steps couldn't have come at a better time—he had grown frustrated with applying practices that simply maintained the status quo, if that.

"This isn't a situation where someone is trying to sell a concept," Fuhrer told me. "It's based on information and education. And as we share that with each other, we've learned how to build that soil back. You can't help but become excited."

The Failed Farmer

On a crisp morning in September, Gabe Brown held a clod in each hand and searched for signs of life—theoretically not a difficult task considering one teaspoon of soil contains more organisms than there are humans in the world. But many of the bacteria and invertebrates that lurk in the dark underbelly of our farm fields exist visually only in the world of high-powered microscopes. So Brown, a compact ball of energy who can somehow combine references to soil biology, farm policy, and animal husbandry in the same sentence, used a less scientific assessment method to compare and contrast the two handfuls—one from his field, the other from a neighbor's.

"When you grab this soil there is no structure," said Brown, referring to his neighbor's soil. Indeed, it had a slabbed, compacted look to it, indicating there wasn't much room for worms and roots to facilitate transfer of water and nutrients. It was also a lighter color than Brown's darker soil, which was the consistency of cottage cheese. "If you have this dark color, you know you have organic matter. I look at it as an investment."

It's an investment in a good crop—just a few feet away stood a field of corn that had emerged from Brown's rich soil, and it was thriving, a rarity in a year when this part of North Dakota had been hit especially hard by drought. But to Brown, that healthy soil represented more than increased bushels in the bin. It was also an investment in his farm's long-term viability and the future of his entire community—human and natural.

During the past decade or so on the 5,400 acres he farms, Brown has put in place an innovative system for building soil health utilizing extremely diverse mixes of cover crops (as many as twenty species at times), no-till cropping, and a type of rotational grazing, called mob grazing, where cattle are put in pasture paddocks for short bursts of intense feeding, often leaving as much as half the plants uneaten so they can feed the soil biome. Brown has more than doubled the organic matter in some of his fields, raising it from less than 2 percent to nearly

6 percent. He has also improved the health of his water cycle, meaning precipitation infiltrates the soil profile instead of running off the surface.

And it's paying off financially. Brown has not relied on commercial fertilizer since 2008, which is significant considering that each 1 percent of organic matter holds the equivalent of $680 in soil nutrients per acre, according to calculations done a few yeras ago by Ohio State University's Extension Service.[6] (Of course, the economic value of organic matter can vary depending on the price commercial fertilizer is going for at any given time.)

These days, Brown's success with building soil health has been so significant that one would be forgiven for thinking he's an anomaly. He's a rock star in the soil health field, and is in big demand as a speaker and YouTube video subject. Walking Brown's farm or viewing one of his PowerPoint presentations can generate a lot of excitement about the potential for linking long-term financial sustainability and soil health. But Brown, whose cherubic face and down-to-earth manner don't quite fit the image of a cutting-edge maverick, is the first to say that he's not special, just, well, more outgoing than most.

"There are people all over doing this. They just don't have the mouth I have," he told me with a laugh while we drove past his crop fields and pastures—many being grazed by cattle—just outside of Bismarck. He added emphatically that all this innovation means little in the bigger picture if farms like his are seen as isolated examples. Brown believes it's also important to remember he didn't attain this level of soil health overnight—it resulted from trial and error. In fact, he's the first to admit that at first it was mostly error.

He and his wife Shelly bought their farm from her parents in 1991, and in 1994 they went 100 percent no-till as a way to save moisture in their cropping system, which at the time produced mostly small grains like wheat. Brown liked the no-till system, but bad weather produced a string of crop failures during the late 1990s. It made things extremely difficult financially, to the point where the Browns were having a hard time borrowing enough money to purchase fertilizer. This forced them to start planting more legumes such as field peas, triticale, and hairy vetch that could fix nitrogen and provide more homegrown fertility while feeding their cattle herd.

"I was using cover crops but I didn't really grasp soil health," recalled Brown. What he did grasp was that his wheat often did better when planted into ground that had just produced a cover crop. His soil's color and texture improved, organic matter levels were rising and precipitation seemed to infiltrate the soil profile, rather than simply run off the surface.

"We had four crop failures in a row, and I tell people today that was absolutely the best thing that could have happened to me and my family, although we didn't think that at the time," Brown said.

I've met a lot of farmers who strike upon a breakthrough way to manage their agricultural enterprise through a combination of trial, error, and in some cases, desperation. What sets Brown apart is that he isn't satisfied to simply put this fresh paradigm in place, making it the new normal for how he manages the land. Perhaps being so close to financial ruin has humbled him enough to realize he doesn't have the ultimate answer. He wants to know *why* he was able to raise organic matter levels and *how* he can raise them even further, for example. What's the next step? Brown is also truly creative in that he doesn't care where he gets his ideas. He recalled with excitement a time when he and Fuhrer were both at a conference and saw a presentation given by a Brazilian scientist about intense multispecies cover cropping systems.

"I turned to Jay and said, 'That's the next step.'"

The Humbled Scientist

There is a photo that has acquired almost legendary status in Burleigh County. It features one of Gabe Brown's fields after thirteen inches of rain fell in twenty-four hours. The picture shows no standing water on this low-lying field, even though plots on neighboring land are inundated. Brown has created a soil ecosystem that allows water to infiltrate quite efficiently. And unlike a field that's been drained through artificial tiling—sending water at rocket speed through the profile and eventually downstream—Brown's fields retain that moisture underground, meaning plants can access it during drier periods. Such a healthy water cycle requires a healthy biological food web.

I first saw this photo in the cramped basement office of Kristine Nichols, who at the time was a soil microbiologist at the U.S. Department of Agriculture's Northern Great Plains Research Laboratory in Mandan, across the Missouri River from Bismarck (she has since gone on to be the lead scientist at the Rodale Institute in Pennsylvania). Nichols, who for many years was on the Burleigh County Soil Health Team, told me this photo is a prime indicator that farmers like Brown are able to increase their soil's organic matter to the point where it is able to, for example, make better use of water. As soil organic matter increases from 1 percent to 3 percent, soil's water holding capacity doubles, she said.

During her early years as a soil scientist, Nichols was taught that a farmer couldn't have a significant positive impact on soil organic matter in a typical lifetime. "We were told this was something we couldn't change, except in a negative way. Now we realize we can change organic matter." That's important, she added, because in the case of organic matter, "You have something that's less than 5 percent of the soil, but it controls 90 percent of the functions."

When Brown and other farmers started seeing positive changes in their soil that weren't supposed to be possible, they approached scientists like Nichols. The microbiologist admitted to me that when she first visited Brown's operation, she wasn't quite sure what to make of what was happening. But for a scientist in a specialized field, Nichols has a refreshing attitude that appeals to practical-minded farmers.

"I'm less concerned about what soil organisms are, and more about what they do," she told me. "We could really learn a lot more about functionality of these organisms."

The first time I met her, Nichols was noticeably energized by the fact that farmers in Burleigh County were sending her "back to the textbooks" when questions came up she'd never confronted before. For example, farmers like Brown seem to be able to raise a good crop of corn with less rainfall than one would expect. Why? Nichols had been poring over plant physiology texts looking for clues. Situations like this make it difficult to determine who is pushing who in terms of cutting-edge innovations in building soil health.

"Just like they challenge me to ask questions, I challenge them," said Nichols. "These guys are so innovative, and they so have the desire for challenge that I don't want them to stop, and I don't want them to allow me to stop. Innovations on the part of farmers are forcing us to come at this from a systems approach and ask deeper questions."

That "dirt" is much more complex than we once thought is becoming increasingly evident as advances in electron microscopes (thanks to medical technology) and DNA testing offer unprecedented glimpses into this fascinating world. But Nichols pointed out that in a way soil is a "big black box" that's just becoming blacker as science unearths new information about what goes on beneath our feet.

"The chemistry happens the way the chemistry happens. But when you throw biology into the mix, it gets complicated," she said while flashing microscopic images of soil organisms on her computer's screen. "In some ways, it's a step backwards—we thought we knew 10 percent of the organisms in soil, now we realize it's less than 1 percent."

But that may not necessarily be a bad thing. It's when farmers begin seeing soil as the heart of an extremely complex and oftentimes mysterious system that they can start taking steps to get at the problem, rather than just treating the symptoms. "We addressed some of the symptoms, which was great, but did we address the bottom line?" Nichols asked.

An example of the bottom line being addressed is when microorganisms do something called "habitat engineering," which has huge implications for not only cutting erosion, but also making sure soil can create its own fertility while staying in place. When soil does not have good aeration and plenty of pore space, it loses its ability to stick together and form strong aggregates. As a result, the water coming in during a rainfall can actually cause weakened soil particles to, in a way, explode.

But soils with more carbon feed themselves, and extra "food" goes into developing a waxy glue—called glomalin—that holds aggregates together, creating a habitat where water can't build up explosive pressure. "They've actually engineered an environment that's safe, that has food, and has the ability to produce carbon to self-perpetuate," Nichols told me. "The more of these aggregates there are, and the larger they are, the less susceptible to erosion the soil is. We've found management can impact this."

Being able to improve soil's ability to engineer its own healthy environment has huge implications on and off the farm. Soil provides at least $1.5 trillion in services worldwide annually, according to the journal *Nature*. For example, soil stockpiles 1,500 gigatons of carbon, more than the Earth's atmosphere and all the plants on the planet.[7] And it's the organic matter that does the heavy lifting: it can hold ten to one thousand times more water and nutrients than the same amount of soil minerals.[8]

In recent decades, great strides have been made in reducing soil erosion to "T," or "tolerable" loss rates—that's the rate at which soil can be lost and still be replaced. This is thanks to conservation tillage and structures such as grassed waterways and terraces. But in recent years, the NRCS has made it clear that to make further reductions in soil erosion, and to in fact start building soil's health and resiliency in the long term, we need to start managing for carbon or "C."[9]

Teaming with Microbes

This goal of "C" means little if the benefits of healthy soil can't be applied on individual farms that put innovative practices in place on a daily basis. Each farmer must apply some big-picture thinking to her or his own situation. A lot of the impetus for the Burleigh County Soil Health Team's approach comes from the popularity of Holistic Management in the region. Developed by Allan Savory over three decades ago, this is a decision-making framework that has helped farmers, ranchers, entrepreneurs, and natural resource managers from around the world achieve a "triple bottom line" of sustainable economic, environmental, and social benefits. This framework is built upon the idea that all human goals are fundamentally dependent upon the proper functioning of the ecosystem processes that support life on this planet: water cycling, energy flow (conversion of solar energy), and biological diversity.[10]

I should point out that Savory has his detractors when it comes to his strident ideas about how livestock can be used to bring back lands hammered by overgrazing and other damaging land uses. Some grassland experts and environmental scientists argue that he overpromises and that many of his claims don't stand up under empirical scientific scrutiny. Savory and his supporters argue that such criticisms are based on outdated ideas of the relationship between animals and the land.[11] My small contribution to this debate is that it is hard to argue with the positive results—many of which I've witnessed firsthand—that have come about on farm and ranch operations that adopt Savory's way of utilizing holism to balance environmental health, economics, and quality of life. The fine details of what farming methods are used within such a system are almost immaterial—it's the big-picture approach that Holistic Management supports which counts.

"The Holistic model has helped get family members and business team members on the same page, helping them all pull in the same direction," Joshua Dukart, a Holistic Management certified educator who has worked with the Burleigh County Soil Health Team, told me. Another important fringe benefit of Holistic Management is that it puts producers in the driver's seat, providing more creative control over what they do out on the land. During a couple of Burleigh County Soil Health Team tours I participated in, I got to see firsthand how such a holistic approach was making itself felt on the land and in the community.

"Take a closer look—anything you tramp down is just carbon in the soil," instructed Fuhrer on a sunny late-summer afternoon during one of these tours. As he said this, he beckoned some 120 farmers and others to follow him into an impressively diverse, chest-high stand of warm season plants: cowpea, soybean, sorghum-sudangrass, pearl millet, graza radish, rapeseed, and sunflower.

This was the first stop on that year's tour, an event that brings farmers, scientists, students, and conservationists from across the Midwest to south-central North Dakota at the end of each summer. As the name of the tour implies, they come to see thriving soil, and the land did not disappoint. Spadesful of fragrant organic matter were unearthed, the results of impressive biological and chemical tests were shared, and crop fields and pastures thriving on that soil were put on display. At one stop at a cornfield, a large jar of water sat next to a six-foot-deep soil profile pit that provided a firsthand look at how deep plant roots penetrate. Suspended at the top of the jar in a wire cage was a fist-sized clump of soil that came from the cornfield. Even though it had been immersed in the water for several hours, the clump was intact and the water remained relatively free of dissolved sediment—a sign that the soil's health was so high that it had been able to engineer its own aggregate stability.

At one point during the tour, Kristine Nichols and Jay Fuhrer used a precipitation simulator to compare the impacts of a one-inch rain on trays of soil representing everything from native rangeland and rotationally grazed pastures to cover cropped no-till fields and conventionally farmed plots. The surface of the conventionally farmed tray was muddied, but little water was in the jar underneath, meaning hardly any moisture was percolating through the compacted soil. A jar that caught surface runoff from the conventional tray was heavy with muddy water, a sign that this sample was prone to erosion. When the conventionally farmed pan was flipped over onto the ground, it was clear only the first half inch or so of soil was wet; the rest was a gray powder.

"You've just made a drought," concluded Fuhrer.

In contrast, the trays growing grasses and cover crops were sending little water running off the surface. Rather, the liquid was percolating down through the soil and dripping into catch jars hanging underneath. It was relatively free of sediment.

"Soil biology is like us—it has to eat," said Fuhrer as he churned up a shovelful of North Dakota earth at another farm stop and held it up for the participants in the tour to see. One way to feed the soil, he explained, is to allow cover crops to be stamped into the ground while cattle are browsing them. This is a pretty radical concept in a line of business that prides itself on taking every last bit of plant material off the land at harvest time—anything less is considered slovenly. That plants can serve an important role as food for microbes and aren't only useful if they can be harvested by machines or animals is just one of the counterintuitive messages emphasized by the Burleigh County Soil Health Team. There are other head-scratchers: planting corn may not always be the best bet financially and agronomically; cattle don't need to spend a long time in grazing paddocks; you don't need as much moisture as you once thought to raise a decent crop; no-till cropping systems alone don't save soil; fields with more varieties of plants, not less, are more resilient in the face of drought.

One farm I visited during a tour was owned by Sanford Williams, who, along with his son Seth, operated a crop and livestock operation. The sixty-eight-acre field that was on display at the Williams place had grown alfalfa during the previous six years. Two months before this particular Soil Health Tour, which was held in early September, the field had been seeded to an eight-species cover crop mix of warm season plants. Timely rains before drought set in during the summer helped produce a vibrant stand, which had resulted in a huge amount of biomass and a build-up of fertility. The Williams family planned on letting their cows calve in the small pasture next to the field, and then turning the animals into the eight-species mix to graze—and stamp biomass. Such a strategy is a long-term investment in the land's, and farm's, overall health—a tough sell at a time when quick applications of fertilizers and chemicals can produce an extremely profitable crop in short order. At the time of this particular tour, corn prices were at record highs.

"I want to plant corn—you can probably guess why," said Sanford while standing in the chest-high mix of cover crops. "Seth wanted to plant cover crops. With crop commodity prices where they are, I'm probably the hard one to convince to do that."

But even the elder Williams conceded that this investment was paying off in ways high corn prices never could—tests showed organic matter and fertility were being built up to impressive levels in the field, all without adding extra fertilizer. Later in the tour the father and son showed off pastures that had been mob grazed. Sanford explained that a lot of his pastures had been full of unpalatable gumweed before.

"Now I can't believe the grass that's growing there," he said. "I'm not a guy who knows his grasses, but I'm seeing species that are producing more feed. But it didn't turn around right away."

Fuhrer backed up that last point by talking about how although diverse cover cropping and mob grazing can rev up the biology of the soil considerably, farmers must take the long view. "We didn't get poor soils in one year and we won't solve this in one year," he told the tour participants.

What was striking about the farmers working on soil health in Burleigh County was that, in a way, doing things in service of microbes had provided them a flexibility not present on conventional farms. At each tour stop, host farmers were invariably asked about future plans for this crop field or that pasture. The majority of them were not set on one concrete choice. They were open-minded—willing to see what nature threw their way before deciding. For example, Seth and Sanford Williams talked about the future of their cover cropped field. After the cattle mob graze it, then what?

"We don't have a definite plan," said Sanford, adding that it depended on how much moisture the area received in the following months—adequate precipitation may mean corn will be a good fit for the field next spring, while droughty conditions could call for a small grain like wheat, which doesn't need as much moisture. Either way they had procured cheap cattle (and microbe) feed from the current stand of cover crops at a time when dry weather had made forage hard to come by. A version of that think-on-your-feet attitude about the next planting season was heard more than once on the tours I participated in.

"It gives you flexibility when dealing with drought," said cattle producer Ron Hein while standing next to a thirty-seven-acre field that used to be all one pasture—in recent years he had broken it up into twenty grazing paddocks. He pointed out that while one paddock is being grazed, nineteen others are resting and rejuvenating, which is particularly important when moisture is short. "It keeps me from having to sell cows."

We also toured the Darrell and Jody Oswald farm near the tiny town of Wing. Using a combination of cover crops, no-till planting, and mob grazing, the organic matter on the Oswald operation has been raised to a respectable 4 percent. Darrell, a long-time cattleman, talked about how working on soil health has made something he never really enjoyed—raising crops—interesting for his family.

"Pretty much everything we do and the decisions we make are based on improving the resource," he said while standing near one of his cornfields, just across the fence from the farm's pastures. "Raising annual crops is exciting for us now."

Such positive energy is infectious and can help attract and keep another important resource, the younger generation, in farming. Gabe Brown and his wife Shelly are thrilled that their son Paul joined the farming operation after finishing college. He's helping perfect their integration of crops and livestock while experimenting with enterprises of his own.

Seth Williams likes machinery and raising crops, skills integral to his family's goal of improving soil health through diversity. After attending a grazing conference, he became convinced animals play a key role in building healthy soil, and he talked his dad into sharing their cattle enterprise with Ron Hein, who is a cousin. Joshua Dukart, the Holistic Management educator, said this kind of teamwork has allowed the Williams and Hein families to concentrate on individual strengths and interests, while contributing to the overall goal of improving the base resource: soil.

"Any given acre, Seth would like to crop it, Sanford would like to hay it, and Ron would like to graze it," quipped Dukart. "But they are able to concentrate on their interests and talents and abilities in certain areas and they're able to complement each other with those. They don't segregate themselves from any other parts of the operation, and [they] still stay very involved with the decision-making as a whole, but basically take the leadership in one area or another."

People involved in the Burleigh County Soil Health Team like to say that if you put soil at the middle, then everything else will follow. It's like handing control over to a powerful, somewhat secretive force that has an outsized impact on all aspects of the environment. And ideally, under the general umbrella of improving the life in our land's basement, everyone upstairs gets a takeaway. In simple terms, Fuhrer and his colleagues can say they are reducing erosion and improving water management, and Nichols gets to see scientific theory and research put into practice while she is given new questions to ponder. But just as important, farmers like Brown who are involved in improving soil health also benefit in some very significant ways. It's a community-based approach to improving a resource that touches on everything from environmental protection and economic viability to the future of rural communities and quality of life.

When I've talked to Fuhrer about progress being made in advancing soil health in Burleigh County and beyond, he's willing to tout statistics on the amount of land growing continuous living cover, the resulting reduction in erosion, and how it's helping reduce input costs and thus increase profits. It's the kind of stuff that looks good in an official progress report or farm magazine feature. But what really excites him is the increasing willingness of farmers to, as he puts it, "speak for the resource." They are not shy about standing before their neighbors and talking about the inherent value of the soil itself—not just what kind of profitable yields it can produce. They are willing to put that resource at the center and work out from there.

And every vibrant community requires a few residents who aren't quite satisfied with the status quo, who keep nudging and questioning even in the midst of success. Brown, for his part, thinks a lot of the practices that have been proven to work on his land will stay limited in scope until farmers learn to observe the land closely, and not rely on cookie-cutter solutions such as chemicals.

"One of the problems I see is a lot of the farmers and ranchers today—and I'll just be blunt—they're disconnected from the land. They oftentimes hire crop consultants, and the farms are so large and the equipment so big they don't get off the tractor and feel the soil and see what's happening," he told me, while holding a handful of his own soil and watching two whitetail deer browsing grass a few hundred feet away.

For Nichols, that kind of close observation and constant questioning can help agriculture, conservation, and science exceed expectations. After all, look what happened when people stopped selling short farmed soil's ability to build organic matter.

"Gabe did something I thought was impossible and instead of telling him, 'Good job,' I said, 'What more can you do?'" Nichols told me during one of our discussions about the team. "I don't know how far we can take it, but I like the idea of not putting limitations or constraints on a system. Can we take it a little further?"

6

Feeding Innovation's Roots

True Believers, Late Adopters, and the Power of the Soil Pit

Michael Werling is, literally, a card-carrying connoisseur of soil health.

"I call it, 'My ticket to a farm tour,'" the northeastern Indiana crop producer said on an overcast August day, referring to the business card he was handing me. We were standing with a couple dozen farmers and agronomists next to a four-foot-deep trench that had been clawed out of a field in preparation for a show-and-tell about the benefits of building soil biology.

The words on the "ticket" left little doubt what was in store for the lucky holder who chose to redeem it: lots of reminders about the precious nature of the land we trod on. Headings at the top said, "My soil is not dirt" and, "My residue is not trash." A third bold line of script across the middle read, "For Healthier Soil and Cleaner Water Cover Crop Your Assets and 'NEVER TILL.'" Buried at the bottom as a bit of an after-thought was Werling's contact information. Given his excitement over the world beneath his feet and how to protect and improve it, maybe it made sense that the farmer's card relegated his address and phone number to footnote status—soil *is* his identity.

While traveling the state of Indiana from one end to the other that August, I ran into a lot of farmers like Werling at workshops and field days dedicated to promoting better soil health on agricultural land. Perhaps that's no surprise, given that events like this tend to attract true believers in the power of healthy humus to do everything from create more resilient fields to clean up water. Agrarians like Werling see wildly successful farming as something that goes beyond what's happening on the surface—it extends deep into that dark world of microbes, invertebrates, roots, and fungi, a place that scientists say is the most diverse ecosystem on Earth.[1] "Dirt Heads" like Werling make supporting and improving that soil universe a priority when choosing what practices to put in place, much like Martin and Loretta Jaus put brushy fencerows and wetlands front and center when making management decisions on their farm.

I was in Indiana because people there had found a way to take the passion of farmers like Werling and use it as an engine for driving change on a whole lot of farms whose owners may not be card-carrying soil sophisticates—they're just looking for ways to cut fertilizer costs and keep regulators off their backs, all the while remaining financially viable. Werling is one of a dozen "hub farmers" located across Indiana who are at the core of one of the most successful soil health initiatives in the country. In just a few short years, a public-private partnership called the Conservation Cropping Systems Initiative (CCSI) helped get 8 percent (around one million acres) of the Hoosier State's crop fields blanketed in rye and other soil-friendly plants throughout the fall, winter, and early spring—times when corn and soybean fields are normally bare. As of this writing, no other Corn Belt state is even close to having that high a percentage of its land protected with cover crops. Indiana's success has farmers, soil scientists, and environmentalists across the country excited about the potential CCSI holds as a national model for bringing our agricultural landscape back to life. Such a model is needed—despite all the buzz these days around providing continuous living cover for the land year-round, little progress has been made in getting a significant number of U.S. farmers to plant cover crops on a regular basis, making what has been accomplished in Indiana even more impressive.

But as I traveled the state, one question dogged me: can such an initiative parlay all this localized interest in one conservation farming technique—in this case cover cropping—into a holistic embrace of a larger wildly successful way of managing soil health? Can the true believers like Michael Werling influence their more industrialized neighbors? Figuring that out could have implications not only for revitalizing the soil universe, but making all aspects of our farming system more resilient via farmer-to-farmer education and support. While touring farms and talking to farmers, soil experts, conservationists, and agribusiness employees, I heard several arguments for why Indiana has succeeded where others haven't. Economics, fear of regulation, teamwork, a tradition of innovation—these reasons and more were brought up repeatedly as I crisscrossed the state. They all play a role. But in the end, a simple love of learning and a feeling of being on the cutting edge of something great may be the ultimate driving force behind making deep changes to a farm.

A Corn Belt Leader

In the early 2000s, around twenty thousand acres of Indiana's farmland was cover cropped, and as recently as 2013, that figure was around half-a-million acres. By 2016 it had doubled to roughly one million acres. That's an impressive growth curve: most people in agriculture agree that cover cropping is a smart practice from an agronomic, economic, and environmental perspective, but frustratingly few farmers have adopted it. One recent estimate is that nationally only around 2 percent of U.S. farmland is consistently cover cropped.[2] That means for some seven months out of the year, most of the Corn Belt's rural landscape is devoid of living plant life, both above and below the surface.

University of Maryland soil ecologist Ray Weil has visited the Midwest numerous times to give presentations, tour farms, and scramble around in soil pits. He recalled a drive he took in Indiana during the winter of 2012. "We must have passed a couple thousand fields and I counted two cover cropped fields," he said. But on one recent return visit to the state, Weil was impressed at how much progress had been made in the intervening years. "Indiana seems to be leading the change. On paper, it doesn't make any sense."

But it does make sense when one takes a closer look at Indiana's intensive team effort to get more of its land growing plants (and roots) for more than a few months out of the year. The Conservation Cropping Systems Initiative consists of federal, state, and local natural resources agencies working with farmers and an array of private businesses, from fertilizer and seed companies to implement dealers. Barry Fisher, a soil health specialist for the U.S. Department of Agriculture Natural Resources Conservation Service (NRCS), said CCSI is rooted in a statewide program that began in 2002 and was focused on promoting and supporting no-till farming (no-till has long been popular among Indiana farmers, given that many of them don't have the luxury of the deep, prairie-based soils their counterparts in Illinois, Iowa, and southern Minnesota cultivate). What he and others discovered during that program's run was that successfully adopting a major change like no-till is more complex than junking the moldboard plow, buying a new planter, and modifying field work schedules. The transition years are critical, especially since a major deterrent to no-till adoption is its reputation for causing a drop in crop yields, particularly the first few years—something farmers call "yield drag." Going cold turkey on tillage may produce conservation benefits on the surface, but the soil underneath is likely to be so biologically dead that it lacks the ability to carry out basic functions like providing nutrients and minerals to plants while making good use of water. "You're going to struggle in any system if your soil fails to function," Fisher told me.

That's when he and other soil conservation experts realized they were going to have to focus on soil health in general, and not just one tool or method, such as no-till. So in 2009 CCSI was born. Under the leadership of Indiana's NRCS, the initiative used federal funding to develop a core group of specialists who were given advanced training in soil health development. Some were even sent to North Dakota's Burleigh County, which has become the model for advancing soil health on farmland utilizing a teamwork approach and where I first witnessed the impressive environmental and economic benefits of bringing soil back to life (chapter 5).

Back in Indiana, these specialists then formed their own regional soil health teams, or "hubs," which consist of farmers, soil and water experts, and Purdue University Extension educators, among others. At the heart of CCSI's work are the workshops and field days it puts on, many of them at working farms. The initiative organizes around sixty such events across the state a year, drawing around 6,500 farmers and certified crop advisers.

Talking about the importance of protecting our soil is nothing new in farming. But the explosion of interest in the biological aspects of soil health in recent years has added a new wrinkle that CCSI has been able to take advantage of. By supercharging that biological activity, farmers can go beyond just putting in a terrace or a grassed waterway to cut surface erosion. They can actually have a positive impact on their entire field's ecosystem using homegrown creativity—an affirming message that they are in the driver's seat.

"That's been a real game changer—the language we use to talk about this stuff," Ryan Stockwell, senior agriculture program manager

for the National Wildlife Federation, told me. Stockwell has been involved in soil health trainings in Indiana, and utilizes cover crops on his
own Wisconsin farm. "Now that you talk about soil structure, all these
benefits from soil health, it creates a lot of excitement. Indiana was just
primed to take advantage of that."

Seeing soil as a living entity that, when fed a balanced diet, can
become self-sufficient, is a pretty big paradigm shift, one that goes counter
to the conventional agricultural wisdom that has dominated society for
over 150 years. The idea that we could use a few select sources of fertilizer
to "feed the plant, not the soil," was popularized by Justus von Liebig, a
nineteenth-century German chemist who is considered the father of the
fertilizer industry. Using research done by, among others, botanist Carl
Sprengel, von Liebig did his best to debunk the "myth" that soil humus
determined the productivity of plants. Rather, he argued, if we simply
focused on, among other things, applying fertility in the form of nitrogen,
phosphorus, and potassium, otherwise known as N-P-K, we would get
exactly what we wanted from plants: big yields. Under such a scenario,
soil was simply a medium for holding up the plant and passing on that
fertilizer to the roots of the crop.[3]

Not every nineteenth-century scientist bought into von Liebig's zeal
for promoting N-P-K to the exclusion of everything else, but the reductionist cat was out of the bag.[4] Ever since, the N-P-K trifecta has become the center of the crop production universe. Not coincidentally,
von Liebig's ideas instantaneously plucked fertility out of the grasp of
farmers, and transformed it into a marketable commodity, one that has
made chemical companies countless billions of dollars over the years.
Once big money gets involved, paradigms, however misguided, become
more entrenched than ever.

The industrial takeover of the nutrient cycle has had a profound impact on the complexion of our landscape and rural communities. Before
the widespread availability of petroleum-based agrichemicals, farmers
needed livestock and diverse cropping systems to return nutrients back
to the land and to control pests. For example, a farmer would raise cattle
on hay, oats, corn, and pasture. The manure from those cattle went back
to the land that produced the feed, and the cycle started over again. The
technologies that went into manufacturing munitions during World
War II were adapted to agrichemical processing. By the 1950s, it was assumed by most farmers that synthetic fertilizers and chemical pesticides
made the fertility-building, pest-disruption abilities of diverse cropping
rotations superfluous.[5]

But CCSI has made progress in getting people to stop viewing soil as merely a plant stand and temporary holder of chemical fertility. I can't count how many farmers I met in Indiana who in one way or another mouthed a version of the phrase, "feed the soil, not the plant." A major focus of CCSI, and its biggest source of success, has been one particular soil health tool: planting cover crops to protect fields during the "off-season" for corn and soybeans. It's a relatively straightforward practice, and sometimes farmers and conservation experts misconstrue cover cropping as the end-all to building soil health. Of course, it's not, any more than no-till was. But it is a handy gateway practice for getting farmers excited about such things as soil bacteria, root interactions, and organic matter. Whenever CCSI team members get a chance, they emphasize that cover cropping is just one tool—albeit an important one. In other words, CCSI isn't just laying out a menu of innovative practices producers can pick and choose from—it's trying to change the very nature of how farmers view soil. "If you can't trigger the 'want-to' in a farmer, all the data won't do any good," Fisher told me. "It's almost an emotional response."

But farmers have to start somewhere on the road to building their soil's biology, and invariably that means experimenting with planting a few acres of small grains or a brassica species like tillage radish as a cover crop. Like soil health initiatives in other states, CCSI has made extensive use of providing government cost share monies to help farmers establish cover crops. But Fisher said his experience with promoting no-till imparted an important lesson about the need for going beyond just subsidizing some seed or equipment.

"If we threw out cost-share money for forty acres and didn't help them in that transition to a new system, they would fail and say, 'I'll never do that again,'" he said, adding that even if the farmer was initially successful, the experimental practice must be sustainable long after the government money is gone.

Customer Support

That's why from day one, CCSI's strategy was to create the same kind of support network farmers enjoy when they pursue more conventional farming practices. That meant not just having government technicians available in each region to help with the basics of bringing the soil back to life. It also requires teaming up with the players that farmers are comfortable working with on a daily basis: fertilizer suppliers, seed dealers, co-ops, crop advisers, and implement companies. I was reminded of how influential such input suppliers are one Thanksgiving while driving through a southwestern Iowa landscape recently left barren by a successful corn and soybean harvest. Rising above the frozen, brown fields on the edge of a farm town was a colorful billboard for a major regional farm services cooperative. In giant white letters on a green background was a message from on high: "Feed The Plant, Not The Soil."

Sarah Carlson, a cover cropping expert for the group Practical Farmers of Iowa, assured me that unfortunately this attitude is the norm within much of the agribusiness community. The thinking from input suppliers is that soil is little more than a place to drive over while applying a few individual products the farmer-customer has purchased from them in order to produce one outcome: high yields. But the CCSI folks are starting to modify this business model on a limited basis. At first it was a bit of a hard sell to get input suppliers on board with promoting cover cropping, since it's a technique that can eventually result in reduced demand for the fertilizer, chemicals, and other products they are in the business of supplying. But in the early years of the initiative, Fisher visited businesses throughout the state and talked about how helping farmers build healthy soils can open up new markets—they need to purchase cover crop seed from someone, for example, and chemical application equipment can be modified to spread seed.

One member of the agribusiness community who early on saw the potential in such a business model was Betsy Bower. She's an agronomist for CERES Solutions, which provides everything from grain handling and agronomic services to fuel and crop insurance to farmers via twenty-two locations, mostly along the western edge of Indiana. She said her company started getting into the cover cropping business around 2010 as a result of customer demand.

"Farmers were coming to us as their trusted adviser and asking, 'What do you think we ought to do? What are the various rates? How do we control weeds?'" she recalled. "As cover crops become more popular, it's going to be in our best interest to learn along with them."

CERES now offers an array of cover cropping services, from soil tests and species selection advice to planting and termination of the plants in the spring to make way for the subsequent cash crop of corn or soybeans. By 2015, cover cropping services made up between 5 percent and 10 percent of the company's business, depending on the branch location. One thing cover cropping does is allow firms like CERES to keep their applicator drivers busy at a time when they would normally be idle or underutilized. They can apply chemicals and fertilizer in the spring, and cover crop seed in the late summer and early fall.

Another key player in CCSI's success is implement companies, which not only sell the planters to put on cover crop seed, but can offer custom field work or modify equipment for seeding. During one field day, I ate lunch with Adam Fennig of Fennig Equipment in Clearwater, Ohio, not far from the Indiana border. He told me that the interest in modifying tillage equipment so that it could plant cover crops "exploded" around 2010. His family's company specializes in mounting cover crop seed boxes, drop tubes, and deflectors on tillage tools. At the time I talked to him, he was doing some sixty modifications per year, mostly in Indiana, Ohio, and Illinois, and the custom enterprise made up about 30 percent to 40 percent of the firm's business.

"But that's about to change in a big way," Fennig said excitedly, estimating that perhaps as much as half of their business would be related to modifying equipment for cover crop seeding by the end of 2017. That's because more farmers are starting to report back major benefits from planting cover crops: everything from reduced soil compaction to yield increases. Many of those reports are emerging firsthand at CCSI field days and workshops.

Fennig said there's a lot of excitement these days around modifying "high-boys" into cover crop seeders. These are the gangly, skinny-wheeled chemical applicators that can drive through standing corn late in the season without damaging the stalks. In a "swords into plough-shares" kind of trick, mechanics in Indiana and surrounding states are modifying high-boys so they can seed cover crops into corn in August, providing a jumpstart on fall growth and providing soil plant protection over the winter. Fennig and Bower credit CCSI for not only supplying them the information they need for providing the proper cover cropping support, but for creating the initial interest in this technique on the part of farmers. "We keep in close contact with Barry Fisher at the NRCS and he lets us know of events in the area," said Fennig. "We try to participate when we can, because Barry can always draw a crowd."

Indeed, the agribusiness support arm of farming was on display at several well-attended CCSI field days I was at. At Moody Farms, a large cropping operation in northeastern Indiana near the Ohio and Michigan borders, seed company representatives showed off an impressive array of miniature cover crop plots containing crimson clover, Austrian winter peas, hairy vetch, radish, rapeseed, turnips, kale, Ethiopian cabbage, sunflowers, annual ryegrass, cereal rye, oats, pearl millet, triticale, and winter barley. As participants walked past each planting, their advantages and disadvantages were described in detail by a local seed dealer. Nearby, Fennig stood next to a tillage implement that had been modified into a cover crop seeder and described how the process works. A shiny red high-boy sat a few yards away and another implement expert described being able to use it to plant cover crop seed in corn that's near the point of full maturity.

The Farmer Next Door

Back in the cavernous Moody Farms machine shed, well-stocked with the tools of a modern row crop operation, some sixty farmers were being reminded that growing corn and soybeans is about more than iron, oil, and chemistry. The reminder came in the form of a question from Dan DeSutter, who raises crops in the west-central part of the state.

"How many of you raise crops with no livestock?" DeSutter said.

The majority of hands in the room shot up.

"So you say," responded DeSutter coolly. "We're all livestock farmers when it comes to soil biology."

DeSutter was one of the first hub farmers recruited by CCSI when it was created in 2009. He and the other hub farmers agreed to host field days and travel to events to talk about their own experiences. CCSI trained them in presentation skills and pays a stipend to cover transportation costs and other expenses. There are also "affiliate" farms that host field days, further helping to tell the soil health story. An added component to the hub concept is that member farms are involved in an ongoing study where information is being collected from their operations on economics, fertilizer use, yields, and, of course, the health of their soil. Beyond that, CCSI collects information from affiliate farms, as well as research plots operated by Purdue University and local Soil and Water Conservation Districts.

The hub farmers represent a wide range of acreage, methods, and growing conditions. DeSutter is on the larger side—he farms five thousand acres near the Illinois border, so he has many of the conditions found throughout the middle of the Corn Belt. Werling, on the other hand, raises 320 acres of corn and soybeans, as well as oats for the local Amish market, on the opposite side of the state near Ohio, putting him more in the eastern Corn Belt.

But no matter where they are located or their size, the hub farmers share a similar passion for improving soil health. To stay connected they usually meet face-to-face for two days every year. The first day is just the farmers; the second day soil experts and agency people are invited to join the discussion. "Somebody starts a topic and it goes onto something else, then those ideas go out to the wider world and other farmers," said Werling. "I love that."

The hub network can serve as a sounding board for proposals that might seem a little "out there" for the conventional ag community, a not-ready-for-prime-time safe place for idea generation, according to Werling. One topic hub farmers have discussed is the idea of seeding cover crops at the same time that nitrogen fertilizer is applied as a side dress during the growing season.

Werling, who has been using a combination of cover cropping and no-till (he calls it "never-till") successfully during the past several years, acknowledges that he is more fixated on the soil biome than the average Indiana farmer. That's why he appreciates the chance to throw new ideas around among a group of people who are as committed to soil health as he is. I asked him, "Like a support group?"

"That's a good way to put it," Werling said with a laugh.

Agents of Change

In some ways, the hub concept is similar to how farm innovations have been germinated and broadcast in farm country for generations. A landmark 1941 study conducted in Greene County in central Iowa traced the adoption of hybrid seed corn during the 1930s. On the face of it, this new technology appeared to be an overnight success—in 1927 it was considered an experimental product not seen outside of college research plots; a decade later it was almost universally planted by Iowa farmers. But through extensive interviews, rural sociologists discovered that the majority of farmers did not accept the innovation immediately, but rather "delayed acceptance for a considerable time after initial contact with innovation."

That's an important point to keep in mind when considering that one can't pick up a farm magazine these days without seeing an article on the importance of soil health—the word on cover cropping and diverse rotations is getting out. However, awareness of an innovation does not always result in immediate adoption—many Iowa farmers who put off planting hybrid seed for years were first made aware of its existence at the same time as their early-adopting neighbors. Although the widespread acceptance of hybrid seed corn over a few years' time is impressive, it's striking that some farmers did not adopt it until a full ten years after their innovative neighbors.[6]

It turns out these early adopters served a key role: they were willing to jump in feet first and test this innovation on their own land almost as soon as they heard about it, and they shared the results with their neighbors in a kind of community laboratory setting. Seed salesman may have been "introductory mechanisms" for hybrid seed, but early adopting farmers were the "activating agents," according to the sociologists. Another important lesson from Greene County is that even after hybrid seed had proven itself on a neighbor's farm, later adopters insisted on experimenting with it personally on just a few acres before making a full conversion.

CCSI's hub farmers are early adopters: people who are trying something new because of a love of innovation and personal goals they've set for their operations. But they don't necessarily have a vested interest in seeing their neighbors make a conversion. "I talk about what I do as a farmer," said Werling of his presentations at workshops and field days. "I don't sell seed. I don't sell fertilizer. I don't work for the government. I think that's an advantage."

Werling's passion for soil health is palpable, and his enthusiasm is contagious as he talks about using crop rotations, no-till, and cover cropping to make even his poorest fields productive. But passion about the soil universe isn't enough, and he knows it. If the majority of Indiana's farmland is going to be planted in continuous living cover, CCSI needs to reach the bigger farmers out there. At one field day Werling attended, the farmers there represented control of some three hundred thousand acres, according to an impromptu survey. When the co-op agronomists and crop advisers attending were included, a total of six hundred thousand acres was represented.

"I don't know if they understand the soil health so much," said Werling of some of the larger farmers. "But there is a lot of excitement over cover crops."

Those bigger operators may not be watching YouTube videos on mycorrhizae fungi—I don't know how many farmers I've talked to who didn't care much about high-speed internet until they became interested in soil health—but we all have to start somewhere, said Fisher. A farmer starts seeing that a cover cropped field requires less nitrogen or yields well in droughty conditions, and then maybe later takes other steps to avoid doing the kind of damage that impedes soil health. What an initiative like CCSI can do is not only support the early adopters out there, but provide an infrastructure for later adopters who are being activated by early examples, and who want to start experimenting on their own farms. Technical expertise, connections with agribusinesses that can provide the seeds and other inputs, cost-share funds to get started on a small scale—these are all offered through the CCSI hub system.

The National Wildlife Federation's Stockwell believes that larger acreage farmers showing up at field days is a sign that CCSI's "saturation coverage" is starting to change the culture. "What the hub farmers do by bombarding farmers from every angle is make it impossible for them to ignore the message," he said. "The middle to late adopters are being reached."

Unearthing Economics

Maybe those later adopters are being reached, but as Greene County's hybrid corn example shows, awareness does not guarantee full acceptance. Fisher said the majority of farmers agree a practice like cover cropping makes conservation sense, but it also has to pencil out financially. That's why the hub farmers were chosen not only for innovative attitudes toward soil management, but also for their ability to track financials and willingness to talk about them. Dan DeSutter, the west-central Indiana farmer, fits the role perfectly. In a sense, farmers like DeSutter serve as an important link in a chain that extends from the ecologically minded motivations of someone like Werling to the more economically centered drivers on Indiana's larger operations. A former financial analyst and commodities broker, he knows how to track trends, talk numbers, and sniff out inefficiencies.

One day while standing in a trench fixing a tile drainage line, DeSutter noticed that roots from the rye cover crop a Purdue University researcher was studying on his family's farm were boring at least four feet deep into the soil. Such "bio-drilling" was impressive, given that over the years the DeSutters had been putting a lot of effort into using a mechanical ripper to break up compaction. Ripping requires a tractor with lots of horsepower and burns lots of fuel.

"That was my aha moment," DeSutter told me. "We were spending all this money on ripping when for a few dollars per acre worth of seed, this plant would be doing it for us. You tell me what's going to do it better: the plant or the steel?"

To DeSutter, that was the "physical" economic argument for building soil health. As he has gotten deeper into cover cropping and talked to other leaders in the field (he traveled to Australia one winter as an Eisenhower Fellow to study soil health building techniques there) DeSutter has also been convinced about the "biological" benefits. Namely, the conventional system of growing corn or soybeans, which covers the land only a few months out of the year with living plants, is actually very inefficient at utilizing all the free sunlight above ground and biological activity below ground.

DeSutter provided a mini soil economics lesson while giving a presentation in the machine shed at Moody Farms. He explained to the gathered farmers that he has doubled his organic matter to 4 percent on many of his acres; as I describe in chapter 5, organic matter is a key element of healthy, biologically active soil. DeSutter then went into a simple calculation showing that the nitrogen he is gaining from this increased organic matter is basically a source of fertility he doesn't have to purchase. "That's like a forty dollar per acre annuity that keeps paying us," he said at one point as the seated farmers took notes.

DeSutter also pointed out that 1 percent of organic matter in the top twelve inches of the soil profile is worth an inch of water storage. "How much is a two-inch rain worth in August?" he asked, following up with an answer in the form of more math. "Let's say it's worth twenty bushels extra per acre. With corn going for four dollars, that's eighty dollars per acre added value. That's resilience."

At another CCSI meeting I attended, central Indiana farmer Jack Maloney talked about how since he started using cover cropping and no-till together, his inputs of nitrogen fertilizer have gone down, but crop yields have continued to increase. He finds cover crops provide fertility to his fields at a more consistent level throughout the growing season—he compared it to a steady sine wave. Applying petroleum-based fertilizer, on the other hand, provides roller-coaster-like peaks and valleys, which don't always match when the crop needs nutrients most. This kind of talk gets a farmer's attention, particularly at times when crop prices are slumping. One corn and soybean farmer working with CCSI, Rodney Rulon, has been taking part in an analysis that shows his use of cover cropping and other methods that build soil health have resulted in a net per-acre value of around eighty dollars, a return on investment of over 320 percent.[7]

Such financial lessons may be directed at conventional farmers, but they are packaged in a way that isn't instantly recognizable to producers who automatically equate the highest yields with large profits. One of the biggest differences between early adopters like Michael Werling and Dan DeSutter and the next wave of farmers who are interested in improving soil health is the role yields play in their decision-making. Werling makes it clear that he does not make a direct connection between high yields and profitability—if he has a few bushels per acre less come fall, that's more than made up for by the fact that he spent less money on inputs as a result of good soil health. DeSutter takes a similar holistic view.

"I don't give a damn about yield," he said in his typical blunt manner during one of our conversations. He then paused and reconsidered. "That's an overstatement. I think there's way too much focus on per-acre yield, and not enough on profit. As a finance guy, I look at what I need to do to make a profit in the long term, to gain a long-term advantage. It's the gift that keeps on giving."

However, the more-bushels-per-acre-automatically-equals-more-profits trap is a hard one to escape. During the CCSI field days and meetings I attended, more than one farmer expressed the goal of getting record-breaking yields while using cover crops. "We've got to get back to science, fellas, if we're going to get to three-hundred-bushel corn," said an Indiana farmer at one point during a CCSI presentation (currently, producing around 170 bushels of corn per acre is considered excellent for the average Indiana farmer). He was half right: science needs to be reintroduced into the soil profile, but so does long-term sustainable profitability.

A Conservation Ethic

One thing that can get lost in all this talk about making soil health pay economically is that for many early adopters the main motivation is care of the land itself. The 2015 Iowa Farm and Rural Life Poll showed that "stewardship ethics" was the most influential factor in farmers' decisions to change how they manage their soil—48 percent of respondents said it was a strong or very strong influence, with economics, at 43 percent, a close second.[8]

Werling, the northeastern Indiana farmer, is acutely aware of the impact his farming activities have on the environment. He farms along the St. Mary's River, which is one of the biggest contributors of phosphorus to Lake Erie. Algal blooms in the lake in 2014 contaminated the water for four hundred thousand people in the Toledo, Ohio, area, forcing a shutdown of the city's drinking water system for three days.[9] "I've been to Toledo Bay," Werling told me. "I'm often the only farmer on those tours. It makes you aware of the algal bloom."

During the CCSI field days I participated in, the often-contentious relationship between production agriculture and water quality hung over the proceedings like a dark cloud. Numerous speakers—whether they be farmers, scientists, or soil experts—made the point that building soil health is one way to be proactive on the issue of protecting the environment and perhaps dodging the hammer of stricter regulation and/or lawsuits.

"I hear you have a million acres of cover crops in this state, and you did that without someone putting a gun to your head," said the University of Maryland's Ray Weil as an opening to his CCSI presentation at a restaurant in southwestern Indiana. Maybe Indiana farmers don't have a gun to their head, but in conversations with me many conceded they felt some sort of stricter water quality regulation of farming practices is inevitable. Watersheds that supply drinking water for the Indianapolis metro area are contaminated with agrichemicals such as the corn herbicide atrazine. "They want someone to pay for it," said hydrologist Robert Barr, referring to Indianapolis officials. Not surprisingly, farmers are working with Barr to show how building soil health can reduce runoff.

An argument could be made for the short-term effectiveness of a top-down approach to cleaning up water when one considers the example of Maryland, where agricultural runoff has decimated fisheries in the Chesapeake Bay. It was determined several years ago that cover crops were the cheapest, most efficient way to capture nutrients before they made it to the Chesapeake, so state officials there instituted a "Flush Tax"—basically a fee all residents hooked up to public water works systems pay. Revenue from that tax is used to pay farmers outright to plant cover crops, usually in the form of a single species such as rye. Maryland farmers can receive as much as seventy-five dollars per acre to plant a cover crop, with other economic incentives thrown in for planting it earlier in the season, among other things.[10]

The result? Around half of Maryland's one million acres of cropland is now regularly cover cropped and agricultural nutrient runoff has been reduced. On the face of it, the program has been a success. But Weil, who has worked with farmers in numerous states, is concerned that most Maryland farmers are narrowly focused on the minimum they can do to adhere to regulations and ways they can qualify for cover crop payments. What happens if economic challenges or shifts in the political winds cause the payment system to be dropped? He prefers what he calls the "rock star farmer" model, where leaders in soil health are driving innovation within their communities.

"The conversation is different in my state, which I think is sad," the scientist admitted to me. "At farmer meetings in Maryland, farmers talk about how they can qualify for higher payments—they don't talk about how they can improve their systems and build soil health."

In a way, such a model is as reductionist as focusing on how much N-P-K is needed to produce a certain yield. When such a narrow view boils soil health down to planting a minimum amount of a single cover crop, it becomes easy to drop that practice once it doesn't pay or it otherwise becomes too big a hassle. The key is for soil health to become the driver of all other farming decisions, rather than one side effect of a few isolated practices. For example, DeSutter has added wheat to his corn-soybean rotation. The small grain long ago fell out of favor in much of the Corn Belt, but since it's harvested earlier than row crops, having it in the rotation gives DeSutter an opportunity to get cover crops planted earlier, providing a jumpstart on winter. Building soil health has to be put on the same level as other farming practices if it's going to weather mercurial markets, shifts in farm policy, or the desire to return to old habits, according to DeSutter. "It's all about priorities," he said. Like wildly successful farmers who base management decisions on how they impact, say, pollinators or grassland songbirds, DeSutter is viewing all his practices through a lens trained on those bugs and biota beneath his feet.

Digging into the Science

I think Weil is onto something when it comes to the power inherent in having a deeper knowledge of something like soil health. When I first started reporting on sustainable agriculture back in the early 1990s, "soil quality" was all the rage within the farm conservation community. Scientists and government conservation officials produced fact sheets and how-to guides on how to protect the soil's ability to remain productive. Most of these methods centered on ways to make conventional cropping systems more "sustainable" by warding off the most destructive elements of this system: intense tillage, overuse of chemicals, compacting the ground with massive machinery.

Some good came out of the soil quality phase—the amount of land that was put under no-till systems that shun moldboard plowing, for example, exploded.[11] But mostly, farmers were not piloting these efforts, and remained unaware of the inner workings of a biologically healthy soil. It didn't have the same connotation as "soil health," which is rooted in building the land's own ability to generate long-term productivity and resilience. It's an ecosystem approach of the most basic kind. As Aldo Leopold wrote: "The most important characteristic of an organism is that capacity for internal self-renewal known as health."[12]

Many farmers undertaking soil-building practices aren't shy about saying nature, specifically the tallgrass prairie biome, is the model they are aspiring to. In Indiana, Wisconsin, Minnesota, North Dakota, and other spots in the Midwest where the most recent soil health revolution has taken hold, what comes up repeatedly from the farmers I talk to—from early adopters to those who are just thinking about it—is how exciting it is to be on the cusp of a new way of looking at farming. An argument could be made that another form of support—an input supplier so to speak—farmers rely on is agricultural and, increasingly, ecological science. And helping farmers unearth what's going on deep in their fields can take them beyond just focusing on one tool like cover cropping. During one series of summer CCSI field days he participated in, Weil repeatedly drove home the point that soil is more than a growth medium for corn, soybeans, and a few small grains or brassicas.

"You can't just throw out cover crop seed and keep doing what you're doing," he said at one point while standing in a four-foot pit that had been backhoed out of a southern Indiana cornfield. As farmers and crop advisers gathered around the trench, Weil used the point of a hunting knife to show where fat corn roots were tracing their way through the profile. Roots were a key part of Weil's lesson that day. It was August 20, and just a few days before, the owners of the field, Clint and Dan Arnholt, had used a high-boy to seed radish and rye into this stand of corn, which was well above everyone's head. Weil estimated there can be a couple hundred pounds of unused nitrogen at the four-foot level, and corn is inefficient at making use of it. Within three or four weeks of being planted, rye and radish roots will be soaking up the excess nitrogen while bringing other nutrients and minerals closer to the surface, he explained.

Fisher and the CCSI team had brought Weil to the state for a week of field days and presentations like this because of his reputation as one of the nation's leading soil ecologists, someone who can put cover

During a field day in northeastern Indiana, soil scientist Ray Weil points out how cover crop roots can extend several feet beneath a field's surface.

cropping in perspective as just one tool for attaining soil health. Soil pits play a major role in such field days. Seeing radish roots bio-drill through what was thought to be an impenetrable soil hardpan caused by years of plowing, wheel traffic, and lack of biological activity can be a real eye-opener.

Michael Werling, the northeastern Indiana crop producer, recalled to me when a soil pit dug in one of his more marginal fields revealed that his use of cover cropping, no-till, and crop rotations had built up the organic matter to the point where a soil expert determined he had slightly modified his soil type. "He said he would have to reclassify the soil," said the farmer proudly while checking out a soil pit at another farm during a CCSI field day. "That's pretty encouraging."

During what were affectionately termed "Ray Days," Weil spent a lot of time in soil pits from one end of the state to the other, talking about the latest innovations in soil science. He should know: besides doing cutting-edge work on the impacts various farming techniques have on soil, Weil is the co-author of the seminal textbook *The Nature and Properties of Soils*.[13] Whether standing in a hole or giving a PowerPoint presentation in a farm's cavernous machine shed, Weil had a consistent message: the science of soil is in flux, and farmers can be on the forefront of this exciting revolution, instead of passive consumers of handed-down knowledge. They can, as has happened in places like North Dakota's Burleigh County, even get ahead of the scientists and be sources of innovation themselves. Weil described how cover crop roots not only go vertical in search of moisture and nutrients, but send branches in a horizontal pattern. The scientist has utilized the same cameras that are used in colonoscopies to trace root channels—it doesn't get any more high-tech than that. Of particular interest to soil scientists these days is the role mycorrhizae fungi can play in building soil health. By interacting with a plant's roots in a symbiotic fashion, such fungi can create a diverse biological universe that's resilient and able to generate its own fertility.[14]

"We're finding out plants send out all sorts of signals underground," said Weil at one field day, citing a recent study that showed older corn hybrids were sending out distress calls when besieged by rootworms; such signals recruit nematodes to attack the pest. "That's pretty cool. That's the way nature works. We didn't really appreciate the role of roots in building soil until relatively recently."

His point, which is reiterated by the soil pits: it's not enough to look at the surface of the soil—take a peek underground as much as possible. In fact, more than once a chagrined Weil admitted to farmers that while revising the latest edition of his textbook he had to re-write the section on organic matter. It turns out farmers can have a bigger influence on their soil's organic matter than scientists like himself once thought. But Weil has another critical message: we don't need to understand the minutiae of how soil protozoa and bacteria interact in order to benefit from this activity on a landscape scale. The key is diversity, which provides the habitat for these interactions to thrive. "If we can encourage the diversity, we can encourage the workings of this system, even if we don't understand all of it," said Weil. "Nature will sort it out."

It's an effective message for the groups that typically attended these field days: a mix of veteran cover croppers and newbies. I talked to several farmers who were having less than stellar results with cover crops, but were hoping they'd learn a trick or two at these events. At each presentation, farmers nodded their heads in agreement with Weil's point that we're all along for a ride on a train pulled by an exciting, if sometimes baffling, ecological engine. This conversation was going way beyond just providing tips on the best seeding rates for rye and radish.

One of the farmers paying close attention to Weil's presentations was Gordon Smiley, who farms 1,200 acres of row crops with his brother

Jeff in southern Indiana. When I met Smiley, the brothers had been using cover crops for a couple of years, and despite a few hiccups along the way, they finally felt this soil-friendly system was an important component of their farm. They have a farrow-to-finish hog operation and the cover crops offer a way to soak up excess nutrients in the manure they apply to their fields. Smiley enthusiastically shared with me how their soil has a crumbly, mellow texture and is full of earthworms.

"What convinced me was the shovel test—digging and seeing the soil underneath," he told me while standing in the shade of a machine shed several yards from a soil pit where Weil had just finished one of his presentations. Up until that point, the brothers had focused mostly on planting a single species like rye as a cover crop, but they were excited to move to the next level of soil health and try cocktail mixes of as many as ten species. Smiley had been watching online soil health videos and attending CCSI field days during the previous few years, and he gave me a long list of websites I should check out as soon as I got back to my computer. Earlier, I had overheard an animated discussion involving him and a couple of other farmers about how one predicts a particular innovation is going to be the next new thing, or just a fad. Smiley was convinced soil health was the former.

"They're way out there," he said of the innovators in soil health he's been observing and interacting with. "We talk about mycorrhizae fungi, we talk about all the bacteria." Then he threw his hands in the air and whirled them around to symbolize lots of activity going on all at once. "It's exciting."

7

Wrapping Around the Wrinkles

Expanding the Land's Potential by Recognizing Its Limitations

We could hear it draining from here," Marge Warthesen told me while sitting in her and her husband Jack's sunny farm kitchen one winter day. The sound they were hearing was a reminder of the vulnerability of the landscape they make a living on. What was draining was a pond just down the hill; it turned out a sinkhole had suddenly materialized beneath it, allowing the region's Swiss-cheese geology, called karst, to reclaim all that surface water in a matter of minutes. A plastic liner was put over the opening the sinkhole had created, and a bulldozer was brought in to excavate soil beneath the pond's dam so it could be used to cover the liner. Jack estimated that as much as ten feet of rich topsoil—topsoil that used to be higher up the slope—was dug up by the bulldozer. This represented decades of land abuse.

He concluded the story with a quip from the bulldozer operator: "Too bad horses can't tip over." In other words, the damage to the land started almost as soon as we hitched draft animals to plows.

The Warthesen farm sits astride a ridge overlooking West Indian Creek, a tributary of the Zumbro River in southeastern Minnesota. From the point where the creek drains into the Zumbro, the larger waterway flows another dozen miles before draining into the Mississippi River. The Warthesen farm is part of the Driftless Area, a series of bluffs, deep valleys, and hills clustered around where the Mississippi touches on southeastern Minnesota, as well as parts of Wisconsin, Iowa, and Illinois. This landscape escaped the leveling effects of the Pleistocene glaciers, so it's dominated by bluffs and narrow valleys. Such a wrinkled effect is beautiful to look at and makes for more surface area per square mile, but it's tough to get around on. Walking some of these sidehills alone is a major undertaking and driving a tractor can be downright dangerous— roll-overs and collapsing sinkholes are all too common. As conservation biologist and historian Curt Meine writes when describing the Driftless: "The corrugated topography does not lend itself to ever-expanding economies of scale."[1] As a result, mega-sized, monocultural agriculture has not gained a widespread foothold in this region.

In fact, when farmers tried a little too hard to bend the region to their desire for large, uniform fields of row crops, it rebelled, leading to horrific erosion in the early part of the twentieth century.[2] Just a few miles from the Warthesen farm in the Whitewater River Valley is a small country cemetery, pretty much all that's left of the town of Beaver, an infamous footnote in soil conservation history. The town was founded in the mid-1850s, and after the surrounding hillsides were stripped of their trees and grass and planted to crops, the Whitewater River became uncontrollable due to all the runoff that resulted. One year alone, the town was swamped more than two dozen times by waters carrying soils loosened from the surrounding hills. Basements were filled with muck and bridges were raised three times in twenty-five years to keep ahead of growing piles of sediment. Finally, less than a century after Beaver's first house was built, the flooding, silt, and mudslides had won—the community was abandoned, and it became known as the "Buried Town of Beaver." Since then, the surrounding hillsides have been replanted with trees and much of the area is in public lands—a state forest, park, and wildlife management area are adjacent to the former Beaver town site.[3]

Given that history, it's no surprise Minnesota's first Soil and Water Conservation District was established in the Whitewater watershed as a way to promote more sustainable land use. In fact, on a wider basis, the region's unforgiving geography prompted the creation of the nation's first comprehensive watershed project. This took place in Coon Valley, which lies in Wisconsin across the Mississippi and south of the Warthesens. Aldo Leopold, along with teams of other University of Wisconsin experts and various government agency personnel, worked with farmers in Coon Valley during the 1930s, helping bring back the kind of diversified agriculture that was more amendable to the realities of the landscape.[4] In short, farmers in areas like Coon Valley and the White-water watershed learned to recognize and respect the Driftless Area's limitations, as well as their own.

Warp and Woof

To pass by a seemingly lifeless stand of grass or trees with Art "Tex" Hawkins is to gain an appreciation for how little one actually sees, even when looking. "It's just bird heaven," Hawkins said excitedly one summer afternoon as he and I, along with Marge and Jack, sat in a van. We had stopped next to a brushy fenceline on the Warthesen farm. A thirty-five-head herd of beef cows was grazing on the other side of the fence. Behind us was corn and hay. The living field border extended both ways and followed the contour of the ridge. Down the hill below the grazing cattle was a pond and beyond that, a thick stand of hardwoods that blocked the view of West Indian Creek.

I had invited Hawkins to the Warthesens' land on a kind of agro-ecological tour to get a set of professional eyes on what appeared to me a good example of a farm that had accepted the limitations of making a living in a landscape that is far from a silent sufferer. Natural resource experts and agronomists have told me that one of the problems with the deep, rich soils that lend themselves to intensive row cropping in flatter parts of the Midwest is that they tend to absorb punishment without showing any outward signs of suffering. That's why the issue of sick soil has caught so many farmers and agricultural scientists by surprise in recent years.

This summer day was significant not only for what was being observed on the Warthesen farm but for who was doing the observing. At first blush, it may seem odd that someone like Hawkins appreciates the benefits that privately held land can contribute to ecological health. For several decades, his employer was the U.S. Fish and Wildlife Service, which is known for, among other things, managing federally owned wildlife refuges across the country. Hawkins was a watershed biologist for the Fish and Wildlife Service in the Upper Mississippi region, and after retiring a few years ago he helped launch a sustainability initiative at Winona State University in Minnesota. Like a growing group of natural resource professionals, Hawkins has learned that refuges do not have impenetrable walls around them. For example, runoff from farms in eastern Minnesota and western Wisconsin is having negative impacts

Marge and Jack Warthesen describe land use in the Zumbro River Valley while wildlife biologist Tex Hawkins (with binoculars) looks on.

on water quality in the Upper Mississippi River National Wildlife and Fish Refuge, a 261-mile stretch of marsh, floodplain forests, and grasslands that begins a few miles from the Warthesen farm.[5]

I first met Hawkins in the early 1990s while following him and a group of farmers through a ridge-top dairy pasture on a midsummer day. It was in the capacity of "bird expert" that he found himself for a time taking part in a multidisciplinary initiative that brought farmers, scientists, and natural resource professionals together as a "Monitoring Team." The Monitoring Team's goal was to develop handy methods for gauging the sustainability of certain farming systems. Thanks in large part to Hawkins, one of the most effective and popular gauges of success adopted by the Monitoring Team farmers involved noting the numbers and diversity of grassland songbirds present in their fields. That's a big deal, considering that monocultural agriculture's hunger for perennial habitat has had major negative impacts on grassland-dependent species such as meadowlarks, dickcissels, and bobolinks, to name a few.[6] Tex is the kind of guy who isn't afraid to contort his mouth into various shapes in order to imitate birdcalls in front of a bunch of cattle producers. Over the years, I've had numerous opportunities to see him lead tours of farms and never fail to be amazed at how he can get a bunch of dairy and beef farmers to suddenly sound like members of the Audubon Society. Once the Monitoring Team farmers took notice of grassland songbird species like bobolinks or meadowlarks, they, in turn, set to figuring out what impacts their production systems were having on nesting success of these birds.[7]

I was impressed with how Hawkins had successfully utilized bird monitoring as the linchpin in a process that led to a group of farms becoming more wildly successful. Perhaps his ability to work with farmers should come as no surprise. Hawkins' late father, Art Sr., was one of Leopold's original graduate students, and, similar to what was done in Coon Creek, worked with farmers in south-central Wisconsin to help renew land and wildlife habitat that had been damaged by decades of highly erosive farming. While working in the area, Art Sr. fell in love with and eventually married a farmer's daughter, Betty. So Tex has long believed strongly that working farms can produce healthy habitat. During many decades of working in rural areas both here and abroad, he has been impressed by operations that are able to draw ecological services from a working landscape. Tex is a particularly big believer in the idea that farms can create an ecological wholeness via what Leopold called, "a certain pepper-and-salt pattern in the warp and woof of the land use fabric."[8]

The Warthesens toughed it out through the agricultural financial crisis of the 1980s, raised four children on the land, and, in general, are proving one can be financially successful while accommodating a little pepper-and-salt in a farm's management menu. I thought their farm would be a good place to get Hawkins' professional assessment of how successful Marge and Jack had been at pretty much letting the land have the final say in terms of how intensely it was managed. Near the abandoned town of Beaver, that philosophy has been put into practice by giving the land back to nature in the form of publicly owned preserves. But farmers like the Warthesens are proving that letting the landscape call the shots doesn't always mean abandoning productive agriculture. In this case, it requires allowing your management style to follow the contours, twists, and turns of the geography—wrapping yourself around the wrinkles, so to speak.

Marge and Jack's farm has always given me the sense of an operation that tucks a lot of ecological efficiency into a tight space—a little bit like the impressive amount of biodiversity that's hidden away in the crooks and crannies of the Driftless Area itself. The home farm is 160 acres (across the road is another 160 acres of family land that the Warthesens farm), but one friend describes it as the "biggest little 160 acres he's ever seen." A drive around the perimeter of the land shows why: the undulating landscape, combined with the diversity of plants, gives visitors the sense that they are entering a different parcel every hundred yards or so. Drive past the cornfield and over a hump in the land, and suddenly there's seven acres of native prairie restored on Conservation Reserve Program ground. Take a walk past a stand of timber and it quickly gives way to a hayfield or a small slough.

While returning to the farmstead from the fields and forests, one gets a grand view of the Warthesens' 3.8 acres of vegetable gardens, which raise enough produce for two local farmers' markets. For several years the gardens also provided produce for a couple dozen local families who belonged to the Many Hands Farm Community Supported Agriculture enterprise Marge operated. She's quick to point out that the garden enterprise has not only produced vegetables, but also its share of intern farmers, some of whom went on to launch their own agricultural operations. Besides beef cattle, the Warthesens also raise lamb, eggs, and poultry for direct sale to consumers. Of the 160 acres, about forty is timber, thirty is pasture, and ten is wetland/wildlife habitat. The rest is crop fields and vegetable gardens.

And the Warthesen farm is virtually one big hillside. Once, Jack showed me a map of the farm and I was struck at how it was basically all one series of contours, with each crescent-shaped field ending in a dead end. It was like a soil conservation textbook illustration of what a contoured farm should look like. The map was tattooed here and there with the initials HEL—Highly Erodible Land—a U.S. Department of Agriculture Natural Resources Conservation Service designation for acres that are so steep that they will lose soil even when row-cropped using a well-managed system. "I'm constantly turning around," Jack said of a typical day of field work on the tractor.

Such a layout makes for plenty of odd corners that don't lend themselves to modern, large-scale row-cropping. The Warthesens' method of managing the land is a combination of proactivity and going with the flow. For example, one crop field in a triangle-shaped area was difficult to turn field equipment around in, so they used USDA Environmental

Quality Incentives Program money to convert it to a rotational grazing paddock for their beef cattle, allowing the feed to be walked out of there in a bovine's belly; four hooves are much nimbler than a lumbering combine.

Even when it doesn't make sense to crop those awkwardly shaped sloughs and corners, most midwestern farms tend to clear them out anyway out of a sense of giving the place a cleaner look. But the presence of the living borders, as well as the timber, sloughs, and other "wild" corners on the Warthesen farm, were not there as a result of neglect, an unwillingness to take the time and effort to "beautify" the farm. Quite the opposite.

Since they started farming this land, the Warthesens have made a conscious effort to combine food production with stewardship. It's been a lot of hard work. Jack grew up across the road from this farm, and remembers as a kid when there were ravines so deep a D6 Caterpillar could work in their bottoms without being seen from ground level. "In places, it was just a huge bunch of ditches," recalled Jack. "You couldn't get across them. You could barely *walk* across them."

Keeping soil on their hillsides and out of West Indian Creek drives a lot of the Warthesens' farming decisions. Over the years, they've broken up large over-grazed pastures and replaced them with a series of smaller ones that are managed using rotational grazing, which helps maintain the health of deep-rooted, soil-friendly grasses while recycling manure. The Warthesens have also replaced contiguous crop fields with contour-hugging strips consisting of diverse plantings of hay, small grains, soybeans, and corn.

The farm utilizes a variety of government initiatives. Besides the USDA's Environmental Quality Incentives Program, Marge and Jack have used the federal Conservation Reserve Program (CRP) to get paid a "rental fee" for keeping their most erosive acres blanketed in perennial vegetation. In addition, trees were planted and managed through a program run by the Minnesota Department of Natural Resources called the Forest Stewardship Program. At one time the Warthesens were also enrolled in the Conservation Stewardship Program, a USDA initiative that paid them for, among other things, putting in food plots for wildlife, reducing fertilizer use, keeping the land in contours, and not applying manure to frozen ground. Their pastures are enrolled in the Grassland Reserve Program, which pays them to keep it in grass. "It's a good program because the right person, or I should say wrong person, would plow up that hillside," said Jack.

The Eyes of an Ecologist

I've been on enough farms to recognize when basic conservation has been set in place. But Tex Hawkins provided a way to go beyond the surface level, so to speak, and sneak a deeper peek into the Warthesens' wildly successful strategies. For example, during our tour the wildlife biologist commented on a shrubby area that was full of game trails. It turned out a boulder was nestled in there somewhere, and Jack grew tired of moving cropping equipment around it, so he let it go back to nature. Marge drove the van close to the tree line above West Indian Creek. Jack pointed out a wildlife planting they established in the early 1990s—maple, black cherry, and pines were thriving. Trees are important to the Warthesen family. When Marge and Jack built a new house on the farm, they included half-a-dozen kinds of wood—oak, cherry, ash, etc.—harvested from the farm, and its dining area features a table made by Jack's father of seven types of wood found on the farm. When planting trees, Jack followed the guidelines in *Landscaping for Wildlife*, a DNR booklet.

Marge parked on a small rise that turned out to be a check dam; in amongst a thick stand of trees and cattails was standing water. Indiangrass above the catchment provided deep-rooted soil protection. Below the dam, a ravine covered with trees plunged sharply toward West Indian Creek, just a quarter-mile downhill. The waterway is a premier trout stream, and the DNR has spent a lot of tax money on it to improve fish habitat.

"This would have delivered a lot of sediment down into the stream," said Hawkins of the ravine. "Now look how it's buffered."

Up the hill was seven acres of CRP set-aside ground that, with the assistance of DNR funds, the Warthesens had planted to oak, ash, and walnut, among other species. Toward the end of the tour, Marge parked the van next to a cornfield, got out, and led the way through a thick stand of hardwoods as a hidden warbling vireo sang its heart out. "They don't even take a breath all summer—sing, sing, sing," said Hawkins with a laugh.

Throughout the entire tour with Hawkins, the Warthesens proudly described how they see bluebirds, tree swallows, meadowlarks, dickcissels, and bobolinks along their fence lines as well as in the pastures. At one point, as if on cue, three adult turkeys emerged from a cornfield a couple hundred yards away on a sidehill and strolled over a hayfield. One of the turkeys was unusually light-colored, almost blond or golden in the July sun.

"God, we got meadowlarks!" Jack exclaimed at one point. They credit rotational grazing with providing good, diverse habitat for these types of grassland birds; research on other midwestern farms backs up this belief.[9] One particularly flashy resident that benefits from the Warthesens' willingness to resist removing a dead tree here and there is the red-headed woodpecker. These birds used to be quite common in the Midwest, where they'd chisel out their homes in dead branches, which they often found in brushy fencerows along pastures. But clean farming has little patience for live trees, let alone dead ones, and North American red-headed woodpecker populations have declined 70 percent since the 1960s.[10] The Warthesens have participated in an Audubon Society initiative called the Red-headed Woodpecker Recovery Program by purposely leaving dead snags and other trees up so the woodpeckers have cavities to nest in. And through the Bluebird Recovery Program, another Audubon initiative, they've mounted nesting boxes on fences.

Hawkins was impressed with how all these separate efforts had been linked to create a contiguous landscape.

"The thing about the field borders and pastures is that they're all connected together with these wonderful fences that are woven together with all these different species of crawling vine as well as low shrubs," he said at one point. "You've got these long strips—they're very narrow, but they're excellent habitat for catbirds, and a lot of other fruit and seed-eating birds like to perch along the fence line. I like the connectivity of the whole situation."

But Marge and Jack's concern for the land goes beneath the surface and beyond their fence lines as well. Next to some old barbed wire, Marge pointed out a small sinkhole, a sign of the karst geology that is a major conduit for contaminants such as agrichemicals, manure, and human sewage to make their way from the surface down to groundwater. That's one reason why rural wells in this part of the state are routinely so contaminated that the water they produce isn't safe for drinking. One estimate is that 20 percent or more of the nitrogen fertilizer applied to farm fields doesn't stay to feed the crop, but rather escapes into the environment. As a result, Minnesota Pollution Control Agency water sampling shows that 70 percent of nitrogen in Minnesota streams is coming from crop fields.[11]

A few yards beyond the sinkhole, Marge stood on the lip of a cliff. Some two hundred feet below was Highway 4, and on the other side of that was an oxbow-kinked West Indian Creek. A neighbor had planted corn in the bend of the river right up to the bank. It was a beautiful view, but also a reminder that no farm is an ecological island. All the efforts the Warthesens are making to improve soil and water quality, as well as wildlife habitat, are dwarfed by landscape-level impacts elsewhere in the watershed.

"There's not much of a buffer strip around that creek where the corn is and in the spring the erosion and the falling off of that field is just atrocious, just atrocious," said Marge as she gazed at the bottomland. "It doesn't take much rain to create havoc down on the Zumbro anymore."

Downstream the Zumbro River empties its contents into the Mississippi, right near the top of the Mississippi River National Wildlife and Fish Refuge. "And at the mouth of the Zumbro you'll see the results—a huge mud plume coming out," said Hawkins. "People are losing a lot of ag ground on the bottoms too, because the river's going crazy and ripping out the banks."

Land isn't eroding because farmers want it to. The reality is that for the Warthesens and their neighbors to stay on the land, they have to make a viable living from it. Rotational grazing can be a low-cost, profitable way to raise livestock. And the recent demand for locally produced foods has been good for the family's vegetable business. But the fact remains that every acre of land planted to trees or grass is an acre not producing corn or some other cash commodity that is easy to market in today's economy, and which is supported by government farm programs. For decades after the harsh conservation lessons of the 1930s, many of those odd corners on farms that were too hilly, wet or otherwise "marginal" to produce a profitable crop on were often safe from the plow—farmers simply avoided them or used them for grazing. But changes over the past decade or so in the federal crop insurance program have now taken much of the economic risk out of farming these rough acres, eliminating an important safeguard for even the smallest smatterings of rural habitat.[12] It turns out government policy does not recognize the limitations of the land.

USDA conservation programs can help relieve the burden of establishing a land-friendly farming system like managed rotational grazing, and the Warthesens have taken advantage of programs that support tree and habitat establishment. But on this summer day, the shadow of farm economics and policy loomed large over their "non-productive" land. At one point, the Warthesens were receiving seventy-eighty dollars an acre annually for renting their CRP ground to the government. A neighbor with land equally as steep as theirs had received over double that amount of cash rent from a crop farmer. When government policy and grain markets determine the value of land, it has repercussions downstream. What goes on in the halls of Congress and at the Chicago Board of Trade has just as much impact on that wildlife refuge as any dam or factory pollution event directly on the river.

Hawkins, mindful that what happens up on this ridge has profound impacts on the river below, noted that one way to make stewardship pay would be through "ecosystem service payments" that would reward farmers for providing such public goods as cleaner water in the watershed. He said such a system has been used in Costa Rica, where he has assisted on conservation projects off and on over the past several decades. "They have a number of different categories of ecosystem service payments that the landowners get," he said. "Costa Rican farmers can receive annual payments to help maintain forest cover, financed in part through European carbon offsets, and this helps sustain clean local drinking water sources, as well as the songbirds that spend their winters in the tropics and raise their young each summer right there on the farm."

"Well, keeping the creek clean would be a public service," said Marge as we climbed back into the van.

It's easy to get down about public policy and market situations that make vulnerable even the least "farmable" parts of that pepper-and-salt landscape. That's why it's important to spend time on a place like the Warthesen farm, where it's clear the siren call of subsidies and even international grain markets falls on deaf ears. The political and economic realities of modern agriculture make being a good steward difficult, but farmers like the Warthesens are highly motivated to work around such barriers. They go to this extra trouble because nurturing the farm's natural habitat is ingrained in their philosophical approach to life and work.

Jack is an avid outdoorsman, the type that finds benefits even in having hollow trees on the place because they provide homes for raccoons, and thus plenty of hunts for hounds. "I've loved wildlife ever since I was little," he told me. A few years ago he wrote a letter to the editor of the *Minnesota Conservation Volunteer*, a DNR magazine, reminding readers that we can do both: farm and support wildlife. One sure sign that the land is truly one with the Warthesens' cultural DNA is the way it sneaks its way into even their small talk. Whether telling a story about coon hunting (a jumper mule used for chasing hounds going down basement stairs as part of a poker bet), ice fishing (a guy getting caught on his own line as he scrambles to turn in a giant carp to win a TV), or deer hunting (a black bear shambling by a tree stand), Jack punctuates the story with "Holy balls!" and a deep laugh. His colorful stories are fueled by the landscape he lives and works on.

Marge's love of the land comes from growing up on a farm ten miles north of Rochester, Minnesota. "I always knew even when I was little I was going to farm. I'd rather be poor on a farm than live in the city," she told me one day while serving up chili made from local ingredients at the kitchen table.

Such an attitude is the difference between living *on* the land, and simply making a living *from* it. When a farm is allowed to become a patchwork quilt of biodiversity, it's simply a more interesting place to spend your days—there's more to it than the county accessor's acreage valuations show. And it doesn't hurt when someone else appreciative of ecological agrarianism comes along once in a while and confirms that all those seemingly unrelated spots of shaggy wildness add up to one smooth unified whole that benefits the wider community.

8

Resiliency vs. Regret

What Domesticated Ag Can Learn from Its Wild Neighbors

The thrum of powerful engines vibrated every fiber in my body as I stood on a hilltop in southwestern Iowa and watched the strafing. Crop-dusters, those sleek pieces of machinery that seem to be all engine and nozzles, were spraying for aphids—thirsty little monsters that can devastate a soybean crop by literally sucking the life juices out of the plant. A lack of timely rains and the absence of aphid predators like lady beetles had made the pests a real economic threat in certain parts of the Midwest that summer. In a year when a late spring had already made a good harvest iffy, no one was taking any chances, even if it meant hiring a daredevil to skim the land while sitting on tanks of toxin. This required putting up with a lot. Besides the noise from dawn till dusk, there were reports of farmers being accidentally sprayed while sitting in their pickups. And then there was an even more nuanced and long-term consequence of all this spraying: it didn't just kill aphids, but any other insects that happened to be in the area, including predators of those aphids.

"The trouble is, it kills the good bugs too," a farmer said to me as he and I watched an airplane drag a train of mist across a creek bottom, its operator creating job security with each squeeze of the nozzle trigger.

Agriculture is full of unintended consequences: examples where one "solution" to a problem is bound to set off a host of unforeseen circumstances, some quite negative. Creating an environment where a pest lacks natural enemies is just one example of industrialized farming's accomplishing more than it bargained for. As a journalist covering environmental issues, I'm self-conscious about reporting only bad news—the default mode for any writer wanting to get their stuff on the front page or to get clicked the most on the internet. But it's hard to ignore all the bad news in our agricultural landscape these days. It's particularly hard to take when one considers that we had made so much progress on issues such as soil erosion up until relatively recently.

As I write this, the Gulf of Mexico's hypoxic "dead zone" shows no sign of abating, and we don't seem much closer to figuring out how to keep pollinators from dying off in massive numbers. In my home state, the Minnesota Pollution Control Agency has found no lakes and only a few streams in Minnesota's southwestern corner safe to swim in.[1] Even the ground beneath our feet is suffering—the journal *Science* has quantified what many conservationists have long suspected: the incredible soil microbial diversity that once dominated the subsurface of native prairie areas has been "almost completely eradicated."[2]

The list goes on. All of these problems can be attributed, in part or in whole, to an industrialized agricultural model that treats the land like a factory pumping out widgets, rather than an interconnected ecosystem. There's nothing particularly insidious about agriculture—any industrial process is bound to come with some hidden and not-so-hidden price tags. And no one in agriculture set out to do harm to the environment. It's almost always a case of unintended consequences, fueled by a reductionist view that you can alter one part of the whole without seeing repercussions elsewhere. As Mary Shelley's Victor Frankenstein said when reflecting on his midnight toils: "with unrelaxed and breathless eagerness, I pursued nature to its hiding places."[3]

But perhaps the most vexing offshoot of unintended consequences is the phenomenon of "regrettable substitution." I first learned about regrettable substitution while reading about how the federal government had banned a type of glue used in manufacturing because it sickened and even killed factory workers. Unfortunately, making that glue illegal led to the proliferation of a replacement adhesive that turned out in many ways to be a worse threat to human health.[4]

Regrettable substitution can be especially—well, regrettable—when it pops up in relation to a practice that is considered particularly innovative. One example of that in agriculture is the no-till cropping system. This is a method that avoids tillage, but rather plants the seed in a narrow slot made in the soil right through dead plant material left over from the previous year's crop. It's an ingenious antidote to the moldboard plow, which opened the Midwest to farming by "peeling back the prairie pelt," as the late agricultural journalist Don Muhm used to say. Plowing created one of the most productive farming regions in the world, but it also exposed the land to incredible erosion while disrupting biological activity and releasing untold amounts of greenhouse gases into the atmosphere. No-till and its various permutations have been utilized in the Midwest since at least the 1970s and took off in the 1990s, when new herbicide systems and advancements in implements made weed control and seed placement into undisturbed soil more efficient than ever. The added benefit to this system is that since it reduces trips across the field that a farmer would usually make to work the soil, no-till saves fuel and time. No wonder around a quarter of all U.S. farm acres are under a no-till system, according to the latest U.S. Department of Agriculture statistics. In addition, hybrid versions of no-till that leave the soil relatively undisturbed but still utilize some tillage—known under the general category of "conservation tillage"—are being used on an additional 20 percent of U.S. crop ground.[5] And it has delivered on its promise: during the 1990s and early 2000s soil erosion on no-till operations plummeted.[6]

On the face of it, no-till is a perfect example of human ingenuity responding to specific problems. We were losing too much soil because of tillage, so we invented a new way to grow plants that did not rely on the practice at fault. Problem solved, right? Not exactly. It turns out no-till comes packaged with a lot of unintended consequences. One of the reasons no-till exploded in popularity in the mid-1990s was that one of genetic engineering's most successful commercial products—herbicide-resistant corn and soybeans—became widely available at that time. Because these crops were engineered to not be killed when sprayed with herbicides like glyphosate, farmers could control weeds without utilizing cultivators and other methods that disturb the soil once corn and soybeans start growing. Cultivating row crops is a time-sensitive practice—it has to be done in that window when plants are high enough to not be buried by disturbed soil, but not so mature that they will be damaged by field equipment making a pass through the rows. It's also very labor intensive—farmers do not miss the countless hours they used to spend under the summer sun driving their tractor up and down rows of corn and soybeans, stopping periodically to clear the miniature "plow shares" that invariably became clogged with clumped soil and weeds. Even more labor intensive was the practice of "walking" soybeans. I assume my skin suffered irreversible sun damage from all those days spent chopping and pulling button weeds, volunteer corn, and cockleburs on our family's farm. Like other farm kids, I was told it built character, but I often wonder if what it did was help me develop more creative ways of complaining bitterly about miserable, seemingly endless work.

It's no accident that the developer of the herbicide-resistant technology, Monsanto, has become over the years a big backer of no-till research. It has even gone so far as to provide farmers financial rewards for adopting their Roundup Ready "package"—herbicide-resistant seeds teamed up with Roundup herbicide. At one time, biotech industry officials talked the U.S. Department of Agriculture into providing farmers a break on their federally subsidized crop insurance premiums if they utilized corn hybrids that were "stacked" with traits like herbicide resistance. By agreeing to this, the government was sending a clear signal to farmers: when you use this technology, you are considered a lower risk to the taxpayer.[7] In a few short years, Roundup became the biggest selling weed killer in the world—in the U.S. alone, well over 90 percent of corn and soybean acres, no-till and conventionally tilled, are raised under some sort of GMO herbicide-resistant system.[8]

Now for the regrettable substitution part of the story. When a cousin of mine planted his first crop of Roundup Ready soybeans, I recall him saying sardonically, "Well, the weeds will be resistant to this stuff in about five years." He was off in his prediction by a few years, but there's no doubt we have a bit of a crisis on our hands in corn and soybean country. Two decades after Roundup Ready technology hit the market, herbicide-resistant weeds are a major issue. Maybe GMO technology can help a crop fend off death by herbicide, but it can't resist the rules of basic Darwinian evolution. As with antibiotic-resistant bacteria, it turns out the more of a particular pest-control agent one uses, the better chance the pests have of spawning offspring that are resistant to that agent. Roundup has basically become a victim of its own popularity. Since it is being used literally everywhere, weeds aren't being thrown curve balls by a mix of treatments.[9] Agronomists are now using online maps to track the growing swath of U.S. farmland where herbicide-resistant water hemp, for example, is taking hold. Another GMO technology—BT corn—has also shown signs of producing hard-to-kill pests, in this case insecticide-resistant corn borers.[10]

One particularly bitter irony of the Roundup Ready technology, and one of the ways it was able to gain government approval, was that it was promoted as a safer way to control weeds. Indeed, the main ingredient, glyphosate, is known as a chemical that is more volatile than other herbicides, which means it doesn't stick around in the environment as long after application, making it less likely to contaminate groundwater, for example. Since the Roundup Ready system allows weeds to be sprayed directly as they grow, the idea is that the glyphosate does its job and then disappears.

This is in sharp contrast to the traditional way of controlling weeds with chemicals—"pre-emergent" herbicides are worked into the soil at planting time, where they ideally can control weeds at that critical period when corn and soybeans are just getting started in the growing season. Because they need to hang around longer to do their job, these pre-emergent herbicides tend to be much more toxic, and much more resistant to being broken down in the environment—a perfect formula for making them a threat to groundwater supplies, for example. It's no accident weed killers consistently show up in water sampling across the Corn Belt.[11]

So the irony of the infestation of weeds resistant to spraying is that one advantage of the Roundup Ready technology—less use of toxic herbicides—is fast becoming a moot point. Farmers are sometimes spraying twice as much Roundup in an attempt to keep ahead of the weeds. Even more troubling, agronomists are recommending that crop producers start re-introducing some of those more toxic herbicides back into their production systems.[12] This isn't just an example of regrettable substitution—it's a case of a regrettable return to the dark ages. For the general public, it means more chemicals in the environment.[13] For the farmer, it's an expensive situation that can harm the very crop they work so hard to put in. I recently drove through Iowa in late June past field after field of soybeans with crispy-looking leaves. It seems a first round of Roundup spraying had left water hemp, a virulent weed, unscathed. As a result, a second round of applications using a more toxic herbicide had been called for. This spraying dealt with the water hemp, but it also made the soybeans look like they'd gotten too close to an open oven door.

And the bad news doesn't stop there: the other unintended consequence of Roundup's popularity is that grassed waterways—strips of perennial vegetation in crop fields that slow water flow, reduce erosion and offer a bit of wildlife habitat—are disappearing. Spray drift from the equipment applying Roundup has doomed them. In addition, scientists are growing increasingly alarmed at indications that glyphosate persists in our soil and water longer than previously thought, increasing human exposure to the ubiquitous weed killer.[14]

No-till farming has another drawback: raising a duo-culture of corn and soybeans without disturbing the soil can lead to a soil profile that lacks the permeable structure needed for healthy plant growth. Heavy implement traffic presses down on the soil, compressing it to the point that air and water can't penetrate. This creates a hardpan so impenetrable that often plant roots can't punch through. One soil scientist described to me a trench dug in a crop field where he could see corn roots hitting such a hardpan and basically taking a right-angle turn in search of moisture and nutrients. Soil experts in Indiana, for example, have explained to me how decades of no-till left the soil so biologically impoverished that crop yields were dropping off. In the ultimate irony, desperate farmers sometimes utilize intrusive ripper implements to cut through the soil profile in an attempt to break up compaction and regain a field's ability to soak up water.

It could be argued that such drawbacks to no-till could be tolerated as long as the technique delivered on its original promise: to reduce soil erosion. And in many cases, it does. But now even that benefit is in danger of being undermined. In many ways, it's not the technology's fault. Just as no-till started to hit its stride, the rules of the game changed. "We have conservation measures that were built for a climate scenario we no longer have," Jerry Hatfield, director of the National Laboratory for Agriculture and the Environment, told me.

Because of climate change, torrential rains have started to become the norm. In addition, decades of chemical-intensive agriculture is taking a toll on the biological health of our soils. This has several consequences, one of them being that soil is losing its aggregate structure—literally its innate ability to hold itself together when exposed to wind and water. As a result, even no-till acres and fields that are relatively flat have been experiencing significant erosion in recent years.[15]

I am struck by the number of Corn Belt farmers I've talked to who are noticing to their great chagrin that the no-till system is not delivering like it should. Like other farmers who have taken great pains to adopt an innovation, no-tillers tend to have the zeal of the newly converted, and so the failure of this system can be a bitter pill to swallow. "None of us who farm want the soil to move—we care," central Iowa no-till farmer Gary Van Ryswyk told me while standing near a waist-high pile of eroded soil. The eroded soil had been collected by researchers and had come off a soybean field he was no-tilling. "I was one of these guys who didn't think we were losing that much soil. I was shocked at how much was being lost."

Lessons from the Land

But a lesson can be learned from no-till's tussle with the rule of regret-table substitution. As I outlined in chapters 5 and 6, farmers have teamed up with scientists and conservationists to return a little biological wildness to their soils, helping gain back some of the benefits no-till had originally produced. These farmers are proving there's nothing wrong with the no-till technique as long as it's being used as a part of a larger holistic system. When farmers and soil conservationists in Indiana began noticing some major drawbacks to avoiding plowing, instead of adding one more chemical or mechanical "solution" to the reductionist mix, they drew from nature's well and looked deep into one of the most diverse ecosystems in existence: the soil universe.

This example shows that insights gleaned from wildly successful farms could help even the most conventional operations avoid the regrettable substitution phenomenon. Because wildly successful farming tends to get at the root of the matter—the soil universe, relationships between plants and animals, etc.—it is more likely to avoid irreversible missteps. That's not to say there won't be unintended consequences—there will be plenty. It just means that when the key elements of a farm's ecosystem health are left intact (or rebuilt), that operation has a better chance of bouncing back from such setbacks.

I once spent an afternoon on Tyler Carlson's central Minnesota farm, where he was attempting to establish a silvopasturing system. Such a technique involves planting trees and other woody species in rows, and grazing cattle and other livestock between them. Once such a system is established, it can produce income from the pasture-raised livestock in the near-term, and from the trees in the long-term. It can also provide significant environmental benefits, including wildlife habitat and the sequestering of greenhouse gases.[16] But as we toured his plantings, Carlson showed me where several young trees had been killed when rodents such as pocket gophers and voles had either fed on the roots or girdled the seedlings. The depredations had been so bad that he had all but given up on having species such as red oak be part of the enterprise.

"Some of the vision of this farm is trying to make agriculture work alongside wildlife," Carlson, who studied ecological restoration in college, said while examining a young white pine tree. It should have been at about the level of our heads or higher, but since its roots had been severed by a gopher, it was dead and listing forlornly to the side in the pasture grass. "But wildlife are pests, in certain circumstances."

However, Carlson's mood brightened when we visited a part of the farm that had a decades-old stand of bur oak, ash, ironwood, elm, and aspen. The farmer had recently cleared invasive buckthorn out of the woodlot, and had then brought in his cattle to graze and keep the understory open. It was starting to resemble a healthy savanna habitat and the cut-back buckthorn was actually serving as a kind of cheap forage for the cattle. Injecting a little bit of naturalness into Carlson's tame pastures had been hit-and-miss, but reversing things and introducing domesticated beasts into a wild corner seemed to be paying off. A farm that has planted every last acre to corn and soybeans doesn't have such an option when things go awry.

We need such flexibility on a bigger scale. In a 2011 paper in the journal *BioScience*, Brenda Lin described how one element of more ecologically based farming, crop diversification, is key if the world's agriculture is to adapt to climate change. For example, diversification is particularly good at fighting pest and pathogen outbreaks, both of which are expected to increase as our climate continues to change. Lin noted that such diversification can be adopted in a variety of forms and on a variety of scales, allowing farmers to choose a strategy that increases resilience while paying off economically.[17]

Agricultural scientists and even policymakers are increasingly seeing the value of diversification. Recently, there's been some fascinating research related to bringing small grains back into the corn-soybean system, adding the kind of diversity that's good for the soil and farm economies. The Kansas-based Land Institute and the University of Minnesota are studying a type of intermediate wheatgrass known as Kernza that can produce human food in the form of grain as well as livestock forage, excellent ground cover, and soil-building root systems—all without being replanted year after year.[18] I've seen Kernza research plots and been on corn-soybean farms that are raising it experimentally, and although a lot of kinks need to be worked out, its potential is exciting.

Someone once asked me if Kernza was a "game changer." By itself, no. But the *idea* behind it—that we need to develop a way to insert into our annual monocropping regimen a little wildness in the form of perenniality—is significant. The fact that Minnesota state legislators have voted to fund such research, albeit modestly, shows that even some midwestern lawmakers realize that the agricultural future does not totally rely on being permanently shackled to corn and soybeans.

Another fascinating study related to diversification took place from 2003 to 2011 on the Marsden Farm, an experimental operation in central Iowa. Three cropping regimes were compared side-by-side. One system was the typical corn-soybean duo-culture. It was then compared to two diversified systems: one involved a rotation where during the third year instead of corn or soybeans a small grain, such as triticale or oats was grown in conjunction with red clover; the other was a four-year rotation of corn, soybeans, small grains, and alfalfa.

The study found some significant energy/environmental benefits from the longer rotations. Synthetic nitrogen use in the diverse rotations dropped 80 to 86 percent, compared to the conventional system. After several years, good weed control was possible in the more diverse systems, even though herbicide use was slashed by 86 to 90 percent. This meant potential herbicide-related freshwater toxicity associated with the diverse rotations was *two hundred times lower* compared to the conventional system. Diverse rotations also required around half the amount of energy use per acre, per year.[19]

These results are pretty much common sense: a greater diversity of plants on the land breaks up pest cycles, helps soil build its own fertility and reduces the need for intense tillage year after year. In addition, legumes like alfalfa and clover help provide for "free" the nitrogen so critical for growing corn. But what is surprising is that the diverse rotations produced competitive yields and similar—in some cases slightly higher—profits compared to their conventional counterparts. This was true during both the transition years and the years when the longer rotations were well established. That's important information for any farmers who are considering making the transition to a more diverse system, but are concerned they can't afford even a year or two of lower profits.

This research makes another important point about profitability: once the diverse systems were established, they were more financially stable from year-to-year. That's because when a system relies less on inputs like petroleum-based fertilizer, it's not as likely to have its bottom line jerked around by price swings in the oil and natural gas markets. If this and other studies show there is more consistent profitability with diversity, why wouldn't a majority of farms adopt such a system? Remember, corn and soybeans are not grown every year when you add small grains and forages to the rotation. That means a farmer needs a way to make something like oats or hay pay during those "off" years when there aren't corn or soybeans available to sell. In most cases, that means having cattle and other livestock present on the farm, or at least on neighboring farms, to add economic value to those plants by using them as feed and to help provide fertility through manure cycling.

The problem is that in many farming communities, livestock have been removed from the land and put into specialized, large-scale concentrated animal feeding operations (CAFOs) while crop farmers focus on just raising corn and soybeans. Geographer Vaclav Smil argues that the Haber-Bosch process, which made it possible to synthesize nitrogen fertilizer on an industrial scale, is the most important technological innovation of the twentieth century. But in *Enriching the Earth*, his history of how Haber-Bosch was developed, Smil expresses regret that the process made it possible to "sever the traditionally tight link between cropping and animal husbandry." This has resulted in a "dysfunctional" nitrogen cycle, Smil argues. "Individual farms, even whole agricultural regions, have ceased to be functional units within which the bulk of crop nutrients used to keep cycling during centuries, or even millennia, of traditional farming," he writes.[20] For example, in areas where CAFOs are particularly concentrated, there simply is not enough land available to spread the liquid manure at rates that the soil can soak it up and utilize it.[21] As a result, at times the waste is applied at double the amount recommended, resulting in significant runoff into local water systems.[22]

The other issue is labor. The Marsden Farm researchers conceded that the diverse systems require a more management-intensive approach, with farmers actually walking the fields, observing changes, and juggling various plant growth schemes, not to mention dealing with livestock. To a specialized corn and soybean producer used to just planting, applying chemicals, and harvesting, this can be a radical paradigm shift, no matter what the profit margin. The outstanding land-based observation skills wildly successful farmers must develop to monitor the ecosystem are simply not as integral to the operation of a monocropping operation.

However, the Marsden study, and others like it, could help make a diverse farming system more attractive to conventional producers by showing that sustainability doesn't require going cold turkey on inputs. It just may require putting chemicals in their proper place—as tools in a toolbox, not the toolbox itself. As the researchers concluded: "More diverse cropping systems can use small amounts of synthetic agrichemical inputs as powerful tools with which to tune, rather than drive, agroecosystem performance."[23]

This is an example of members of the mainstream agricultural science community recognizing the benefits of injecting a little wildness into conventional farming as our growing conditions become less predictable. When optimal growing conditions—just enough rain, but not too much, for example—aren't present, conventionally produced crops tend to falter. It turns out those thoroughbred hybrids often require thoroughbred-level growing conditions, something that's increasingly hard to come by in this brave new world of a volatile climate. To borrow a horse-racing term, we increasingly need mudders on the track as less than ideal conditions became the new normal.

Every little bit helps. A major survey conducted by the U.S. Department of Agriculture in the Upper Mississippi River watershed showed that during the drought that baked much of the Midwest in 2012, farmers who utilized cover cropping to protect the soil before and after the regular corn-soybean growing season preserved enough precious moisture to provide a yield bump for the cash crop that was planted the following spring. In the case of corn, cover cropping provided a yield advantage of around eleven bushels per acre.[24]

Stripping Conservation to the Bare Essentials

Remember how farmers like Marge and Jack Warthesen (chapter 7) have decided for various reasons to share portions of their property with nature by leaving and establishing stretches of wildlife habitat? It turns out you don't have to have a degree in wildlife biology or have a dog-eared copy of the *Sibley Field Guide to Birds* to see the benefits of native prairies, windbreaks, and even wetlands. The Wild Farm Alliance has documented numerous examples of farmers utilizing beneficial species to keep insect pests under control, for example.[25] Just as Aldo Leopold realized that wolves are an important link in maintaining healthy timberlands, research around the world shows that well-balanced predator-prey relationships can play a key role in sustainable food production.

For example, a *Proceedings of the National Academy of Sciences* study published in 2015 reported that bats boosted corn yields on Illinois farms by 1.4 percent, which equated to a $1 billion ecological service worldwide to corn producers. It turns out the bats are voracious eaters of corn earworm moths, whose larvae can do major damage to corn. The presence of earworm larvae was almost 60 percent higher when bats were excluded.[26] A 2014 University of California-Berkeley study showed that hedgerows of native flowering shrubs planted along crop fields helped keep pests under control by providing habitat for natural predators such as parasitoid wasps.[27] In fact, in the wake of vegetable contamination scares in places like California, food companies and regulators put pressure on farmers to remove any wildlife habitat adjacent to produce operations. The thinking was that feces from roaming wild animals was the source of illnesses caused by E. coli and salmonella. Follow-up studies have discounted those fears and even shown evidence that wildlife habitat adjacent to vegetable plots can help produce a kind of disease buffer, reducing the presence of illnesses that originate from wildlife as well as domesticated livestock.[28]

On a June day in central Iowa, I saw a dramatic example of how conventional agriculture can benefit when a little wildness is allowed to creep in through the cracks in domestication. I was standing next to Gary Van Ryswyk, the no-till farmer, and that troubling pile of eroded soil, which had been excavated from a flume-like device that channeled water flow through a system for measuring runoff. A shovel had been jabbed into the pile by a graduate student as a reminder that this wouldn't be the last time the measuring device would have to be cleaned out. But the researchers Van Ryswyk was working with were somewhat surprised at the almost total lack of eroded soil being collected by a flume just a few hundred feet away from where we were standing. The soybeans above that particular collector were also being grown under a no-till system by Van Ryswyk, and the field slope was the same. The difference was that growing in strategic spots on the second field plot were patches of native prairie.

Back in the early 2000s, a group of scientists and others in Iowa began discussing ways of using perennial plant systems like prairie to add some diversity to fields dominated by annual row crops like corn and soybeans. These people knew better than to propose replacing whole fields with perennial habitat. Instead, the idea was to plant thirty-to-fifty-foot-wide strips of prairie in strategic locations throughout crop fields. Out of this has emerged STRIPS, which stands for Science-based Trials of Rowcrops Integrated with Prairie Strips. Fostered by Iowa State University's Leopold Center for Sustainable Agriculture, the research project was launched in 2007 at the Neal Smith National Wildlife Refuge in central Iowa. This is an excellent place for such a study. For one thing, the topography provides a real challenge to keeping soil in place—slopes of 6 percent to 8 percent are not uncommon. In addition, the wildlife refuge is undergoing a long-term transition from crop fields to native re-stored prairie. Van Ryswyk already happened to be raising crops utilizing no-till on refuge land awaiting transition, so he agreed to take part in the study.

Scientists involved with the project told me during my first visit there in 2011 that when the research started out, the assumption was that planting native prairie in crop fields would provide some environmental benefits such as increased habitat for wildlife and pollinators. They also assumed the strips would slow overland water flow, allowing it to better soak in and thus reduce the amount of soil that made its way to the bottom of these hills, and eventually into the wider watershed. Indeed, there has been an increase in the number of grassland bird species using the studied crop fields, and yes, pollinators are doing well in the strips. But researchers weren't quite prepared for just *how* successful the strips turned out to be at slowing down water and cutting erosion.

It turns out planting 10 to 20 percent of a crop field to native prairie consistently cut erosion by an astounding 95 percent. The plantings reduced phosphorus and nitrogen runoff by 85 to 90 percent.[29] "It's hard to improve on 95 percent," Matt Helmers, an ISU engineer and one of the STRIPS coordinators, said to me as we walked through one waist-high micro-prairie in the midst of a soybean field.

In a way, this research was in direct response to the shortcomings emerging from the use of no-till. Van Ryswyk and several farmers in this part of Iowa are pioneers in this technique, and had expressed frustration that it doesn't seem to be holding up in the face of the increasing number of heavy downpours they're getting in the region. ISU agronomist Matt Liebman told me that at times the rains can be so intense they simply lift plant residue off no-tilled fields and float it away, leaving the soil completely vulnerable.

"No-till is no longer enough," he concluded.

Planting patches or strips of grass in fields to reduce erosion and overland water flow is nothing new. However, such plantings are often dominated by a monoculture of, for example, bromegrass, which tends to lay down in the face of extreme flow, allowing the water to race downhill. What differentiates the STRIPS research is the use of native prairie. It seems the thick stems of prairie plants are effective at slowing water, and anything along for the ride, via a kind of pinball effect. Such zigzagging can help dissipate the water's energy, and it eventually slows enough to drop any soil it's carrying and soak into the soil profile. As a result, planted into the strips were stiff-stemmed warm season grasses such as Indiangrass, big bluestem, and little bluestem, along with a wide range of erect forbs such as asters, bee balm, blazing star, bush clover, coneflower, goldenrod, and native sunflower.

The tallgrass prairie biome once blanketed 85 percent of Iowa, and its replacement by annual crops is considered the most substantial transformation of any major ecosystem in North America.[30] But when environmentalists suggest returning more prairie grasses and other perennials to the landscape, reaction from the agricultural community is often out-right hostile. Farmers envision a "Buffalo Commons" scenario where row crops are replaced with thousands of square miles of nature preserves—leaving no room for food production, or the people involved with it.[31]

What's so exciting about the STRIPS research is that it proves the value of focusing conservation on critical areas that are relatively small, providing outsized environmental benefits in regions where working farmland is a key part of the economy. This fits with the idea of using targeted or "prescriptive" conservation to reach environmental goals, which remain frustratingly out of reach in farm country. A 2015 study published in the *Journal of the American Water Resources Association* outlined the limitations of traditional farmland conservation strategies, which, through such agencies as the U.S. Department of Agriculture Natural Resources Conservation Service, often consist of getting practices applied to as many acres as possible, regardless of the ultimate impact. The study looked at conservation practices in Minnesota, Wisconsin, Iowa, Illinois, and other states that make up the Upper Mississippi River–Ohio River Basin—the traditional Corn Belt—and found that applying water-friendly techniques such as cover cropping and reduced fertilizer applications across the entire region won't do enough to reach water quality goals. But by targeting areas that are particularly vulnerable to runoff with a mix of intense conservation techniques—everything from utilizing wood-chip bioreactors near tile drainage lines to restoring wetlands—nitrogen runoff in the Upper Mississippi–Ohio River Basin could be reduced by 45 percent. That percentage is the threshold the Environmental Protection Agency has set for reducing the size of the dead zone in the Gulf of Mexico.[32]

"We often pay for practices rather than outcomes," ISU's Liebman said during one of our many conversations about the prairie strips research and farm policy. "It's not the best use of society's dollars. It's clear that random acts of conservation are a waste of money."

In a sense, a prairie strip is the opposite of being a random act of conservation, given that it occupies a small percentage of a field, and that its location is pinpointed to produce the biggest bang for the runoff-stopping buck. The use of remote sensing technologies like LiDAR (light detection and ranging) provide efficient ways to determine *exactly* where runoff and erosion are likely to occur in a given field.

On my most recent visit to the STRIPS research site, the team was in the midst of transferring this tool from the wildlife refuge test plots to working farms. The researchers had brought farmers to see the strips firsthand and conducted extensive outreach in rural areas to spread the good news. Numbers were being crunched to determine how much establishing a prairie strip would cost a farmer, and whether current government conservation programs could help cover at least part of that bill. The STRIPS team was also developing a system for helping local conservation technicians design strips for farmers they were working with. At last count, over two dozen farmers in and around Iowa were planting their own versions of prairie strips.[33]

But prairie strips are no silver bullet, and we've already seen the dangers of relying too heavily on a conservation technique that generates a lot of initial excitement. For one thing, the strips keep soil from leaving a field and making its way into waterways, but erosion *within* fields still can occur under such a system. Helmers and Liebman explained to me that a field that was environmentally vulnerable because of its steepness needed a systems approach to protect it from the top all the way to the bottom. Techniques such as cover cropping and no-till production could be used to provide overall protection, with the prairie strips serving as a kind of conservation "polisher." Now that the innovation has left the test plot, farmers are putting their own creative spin on making the prairie strips more effective, as well as economically viable. For example, one farmer in southwestern Iowa has grazed cattle on the prairie strips he established with Iowa State's help, providing cheap forage for his operation while building soil and perennial plant health in a way that reduces runoff even more.

While touring the wildlife refuge, it occurred to me that one potential barrier to adoption of this system is that, as with many conservation measures, the prairie strips have the potential to produce more benefits off the farm—cleaner water, for example—than on. Van Ryswyk argued that even the most conscientious farmers aren't likely to notice the difference on their land, given that much of erosion's impacts can creep by unnoticed. In other words, prairie strips in row-cropped fields are mostly a public good, and getting them established may require public support to prime the pump.

Van Ryswyk said it also wouldn't hurt if more farmers were aware of how much runoff occurs in even well-managed fields. "One of the big barriers is, like me, most farmers truly believe they aren't losing as much soil as they really are," he said. Van Ryswyk's comments were a reality check. Here was someone who cared deeply about the state of his soil, but it took him being involved with an intense research project to truly see that the practices he was using weren't as effective as he thought. Not every farmer has that opportunity.

Perhaps before a technique like prairie stripping is recognized as a public good, benefits that go beyond reducing erosion will need to be identified and promoted. I thought about that while walking through one of the contoured prairies with Pauline Drobney, a U.S. Fish and Wildlife Service prairie biologist who works on the STRIPS initiative. While patches of native grasses and forbs interspersed with corn or soybeans are not as optimal as having vast tracts of wild grasslands extending to the horizon, she was excited about the potential for prairie strips to provide various ecosystem services in the heart of row crop country. For example, numerous pollinators use the strips, including over thirty native bee species. They have also proven to be important habitat for birds, including several grassland species that have experienced extreme population declines as row crops gobble up pastures, meadows, and grassy corners on farms. Sometimes it doesn't take much: a study in Kansas showed over twice as many total number of wild bees and almost 50 percent more bee species were found along highways planted to native prairie when compared with weedy byways.[34]

"Imagine a landscape where you have prairie plantings like this interspersed," said Drobney as soybeans grew above and below us, while dickcissels sang their liquid song from grassy perches and bees puttered between flowers. At that moment, the terms "pepper-and-salt pattern" and "warp and woof of the land" popped into my head.

9

Which Came First,
the Farmer or the Ecologist?

The New Agrarians and
Their Environmental Roots

When Peter Allen was pursuing a doctorate in restoration ecology at the University of Wisconsin, his view of the natural world was relatively straightforward: the best way to build a healthy ecosystem was to keep humans out of the picture. Before starting the PhD, Allen had this worldview reinforced while obtaining a master's degree from UW in conservation biology and sustainable development and before that an undergraduate degree in environmental science from Indiana University. Allen was a vegetarian at the time, and the role of livestock in devastating the land played a particularly large role in his academic-based ecological worldview.

"I was coming from a traditional ecological restoration ecology perspective of humans are kind of bad on the landscape and we're trying to restore some sort of native natural balance," Allen told me on a beastly hot summer day. As he said this, we were standing on the upper slopes of a steep hillside overlooking southwestern Wisconsin's Kickapoo River Valley. We had just hiked past his small herd of beef cattle grazing on the lower parts of the hill. Before that, we had toured other pastures, as well as a restored prairie habitat on his farm, while riding around in a solar-powered Polaris Ranger. Allen had stopped periodically to point

Southwestern Wisconsin farmer Peter Allen is using livestock to refurbish oak savanna habitat on his land.

out the other livestock he and his wife Maureen were raising on this farm—hogs, sheep, and goats—and which they sold to consumers looking for naturally raised meat.

To say Allen had modified his view of how to have a positive impact on the landscape is an understatement.

It turns out a surprising number of people with academic and professional backgrounds in natural resources fields are returning to the land as farmers, rather than as wildlife refuge managers, conservation officers, or ecological scientists. No official numbers are available, but interviews I've conducted over the past few decades show getting an agronomy or ag business degree from a land grant university—the traditional academic-based pathway into production agriculture—no longer monopolizes the way young people enter farming. A striking number of people are going into food production after receiving training and working in the fields of environmental science, wildlife biology, ecological restoration, and other areas related to protecting and studying the environment. These are people who went into the field set on the belief that by working for a natural resource agency or an environmental nonprofit, they could help leave the land better than they found it. But somewhere along the way, they took an off-ramp and dived into a profession many environmentalists see as the antithesis to a healthy ecosystem: agriculture. When I ask them why, the answer is invariably a variation on a theme: "I felt I could have a bigger impact on the ecological health of the land through farming."

They aren't buying into the narrative that farming and a healthy ecosystem don't mix. In fact, as a result of advances in sustainable agriculture management and innovations related to everything from grass-based livestock production to cropping systems that build soil health, there's a new generation of ecological agrarians who feel producing food and cultivating a healthy natural landscape go hand-in-hand. This is particularly true in a place like the Midwest, a region of the country where private acres in agricultural production is the dominant land use. National parks, wildlife refuges, and other publicly owned natural areas are few and far between. Want to have a positive impact on your eco-zone? Then figure out how to inject a little wildness into food production.

This may come as a surprise to other environmentalists and even the average citizen who reads all the dire agro-environmental headlines out there, but it's a comment on just how far sustainable agriculture has come. And from what I've witnessed, these are no back-to-the-land, living-as-peasants fantasies these people are acting out. They are working to prove that wildly successful farming can be economically viable.

So, which came first, the farmer or the ecologist? As the farmers featured in this chapter are proving, that may be a moot point. Once they get established on the land, they are proving that the same skills that make them good ecologists are useful in generating a living on the land. As one of these farmers, Bryan Simon, told me, that shouldn't be such a surprise.

"Ecology is all about seeing the big picture and not focusing on only one aspect," he said. "Agriculturalists that are able to look beyond simply the number of bushels produced per acre and take a more holistic approach will be more successful and resilient in the long run."

Redefining Agriculture

As we stood on that Wisconsin hillside, watching a curtain of rain march down the Kickapoo Valley, Peter Allen ticked off all the amenities that drew him to this 220-acre piece of ground. There is remnant savanna habitat, which, with its mix of oak trees and grasslands, is one of rarest natural habitats left in the Midwest. And there are also two springs and a wetland, which provide water for the farm as well as wildlife. Then there are the remnant native plants growing in the hillside pastures—as we zigzagged up the hill Allen pointed out bee balm, goldenrod, and black-eyed Susan.

"It's got all the various ecotypes of the Driftless Area represented in this *one* place," Allen told me excitedly, sounding like someone who had spent years studying ecology—which he had.

But then he looked down at what at first glance seemed to be the least interesting part of this farm, and got *really* animated. Below us, tucked into a pocket between the hillside and a trout stream called Camp Creek, was something all too common in the rest of the Midwest: a forty-acre field of soybeans. It looked flat from our perspective, but Allen explained that when heavy rains hit the farm, a surprising amount of the field's soil washes into the stream. Any contaminants along for the ride in that runoff eventually make their way to the Kickapoo, which connects to the Wisconsin River and eventually the Mississippi.

This domesticated, monocultural reminder of industrial agriculture's dominance of the landscape stands out in stark contrast to the naturalness of the rest of Allen's ecological oasis, which he calls Mastodon Valley Farm. And he is thrilled to have the field there. It will provide him a prime opportunity to put into practice years of classroom training, reading, research, and, most recently, on-the-ground experiments. He can't wait to begin the process of converting the field to prairie, and eventually making it part of the rotational grazing system he has set up for the livestock being produced here.

Allen's change of heart on how restoration ecology can be executed on the midwestern landscape is traced to some of what he studied while in college. Specifically, he had researched the ecology and history of the oak savanna ecosystem, which consists of hardwood trees like oaks interspersed with tallgrass prairies (bur oaks do particularly well in savannas, since their thick bark protects them from the effects of fire). In effect, oak savannas are the transition between prairie and the woodland, so you have the best of both worlds. These are highly diverse habitats because of this mix of trees and grasslands. One estimate is that by the time Europeans arrived, roughly fifty million acres of oak savanna habitat existed in a band stretching along the eastern edge of the Great Plains from Texas into southern Canada. There were also scatterings of this habitat in Indiana, Michigan, and Ohio. Most of the oak savanna habitat was wiped out to make way for farming in the latter half of the nineteenth century. At best, thirty thousand acres of the habitat remains in the Midwest today, with each parcel amounting to less than one hundred acres.[1] However, because of the difficulty of row-cropping some of the steeper hillsides that make up regions like the Driftless Area of southwestern Wisconsin, southeastern Minnesota, northeastern Iowa, and extreme northwestern Illinois, this region has prime pockets of oak savanna habitat remaining. In fact, scientists believe this region has the largest area of what they call "restorable" oak savanna.[2]

What Allen came to realize was that oak savannas are not a climax community—what, when left to its own devices, nature will aspire to. When the first European settlers arrived in regions like the Driftless, what they should have found was closed-canopy forests. But journals made repeated references to oak savannas, a habitat reliant on intervention.

"If you leave land alone, it doesn't just turn into savanna," said Allen. "It takes quite a bit of active management."

It's now commonly believed that Native American societies—with varying degrees of intent—managed these habitats using fire. The result was a habitat rich in herbivores like deer, elk, and bison, and which produced numerous nuts and fruits, including acorns, hazelnuts, prairie crab apples, plums, strawberries, raspberries, blackberries, pawpaws, hawthorn haws, gooseberries, and highbush cranberries. In other words, the Native Americans' management system was helping the land produce more food for human consumption. Just because the first European settlers didn't recognize what they considered an agricultural landscape—squared off fields with fences and monocrops of grains—didn't mean it wasn't being managed to produce human sustenance.

Oak savannas also have the potential to produce lots of wildlife habitat, as well as to sequester greenhouse gases—the trees trap and store carbon for a hundred years or more, until they die, burn, or are cut down. The grasses, if they are managed well, can keep sequestering carbon in a continuing cycle (there is not yet scientific consensus on just *how much* carbon grasslands can sequester long into the future).

"Grasses and trees are duking it out in this never-ending battle," said Allen. "And sometimes it's going to be all grass, and sometimes it's going to be all trees. Sometimes it will be a mix of both."

While working on his PhD dissertation he developed a model of a farm patterned on oak savannas—it would integrate a polyculture of tree crops, fruit and nut production, as well as multiple species of livestock that are rotationally grazed. The animals would take the place of fire as a way to maintain open spaces between the trees. Such a management system would require actually removing trees to create that open space, an idea that Allen concedes "freaks out" his environmentalist friends. At about that time, Allen was in a bar talking about this idea with a friend who was into permaculture—a method of food production that relies on perennial species that don't have to be replanted every year. "He was like, 'Oh yeah, that's kind of like what Mark Shepard does,'" Allen recalled the friend saying.

Shepard's New Forest Farm near Viola in southwestern Wisconsin has become a model for integrating, or "stacking," various enterprises utilizing permaculture plants such as fruit and nut trees. Allen ended up going to New Forest Farm to collect data for a case study and, "All of a sudden this idea for a dissertation wasn't just an idea in my crazy head— there was an example of it on the ground."

In 2012, he pitched a tent on New Forest Farm with plans to spend a week. Allen was so struck by Shepard's use of the agricultural savanna concept that he ended up spending an entire summer there with Maureen, who has a degree in zoology. They extended their stay through the winter by living in a shack on the farm, and Allen convinced some friends to go in on helping to buy six steers. By their second summer on New Forest Farm Peter and Maureen were grazing twenty head of cattle, as well as pigs, sheep, and poultry, among the hazelnuts and other woody species growing there. Allen was hooked. He was also convinced that he didn't need to get his doctorate to accomplish his goal of using agriculture to restore oak savanna ecosystems. After spending a decade in graduate school, he dropped out six months shy of finishing. He and Maureen began looking for land that had that right mix of natural habitat and cropped land, and in 2014 bought what has become Mastodon Valley Farm, eight miles from New Forest Farm.

Allen's decision to leave school was opposed by just about everyone he knew, including Shepard. But he felt putting off getting established on the land would cost him precious momentum—he was finished reading about ecological restoration, now was the time to put it into action. If he hadn't jumped at this chance, "I would probably have a job teaching, talking about starting a farm someday, always talking about starting a farm," said Allen. "Even when I was in the academy, I was always an outlier. I never quite fit in. Most of the ecologists thought I was crazy, because I'd talk about bringing in goats to manage invasive species, or we should bring cattle in to manage these grasslands. And then the ag people thought I was crazy because I was thinking about diversity and grassland birds and pollinators and these kinds of things."

When I visited Mastodon Valley Farm, Peter and Maureen were starting to wrap up their third growing season on the land. A lot had happened in a short time. They had built a cabin and were pretty much living off the grid, utilizing solar power and having to drive some eleven miles to utilize an internet connection. The livestock herd that was launched on New Forest Farm had been transferred to rotationally grazed pastures on the Allen place. They also started a meat marketing enterprise to help make their ecological restoration project economically viable. Like many of the ecological agrarians I've met, Allen is hoping to get the environmentally conscious consumer to financially support this method of managing the landscape.

"I talked to a guy once and he's like, 'Well, I'm not a farmer, I don't have land. There's nothing I can do to influence the landscape.' I said, 'You influence the landscape three times a day. Every time you put something in your mouth you are influencing a piece of ground,'" Allen recalled.

Mastodon Valley Farm sells its meat utilizing the Community Supported Agriculture (CSA) model. Consumers in Madison and La Crosse buy a "share" in the farm, which entitles them to a monthly delivery of beef, pork, and lamb. By their third season, Peter and Maureen had about sixty regular customers and were marketing roughly twenty beef cattle, twenty-five hogs, and twenty to twenty-five lambs annually using this model. Allen feels if they could double that amount it would fit their economic needs, but would still not overtax their land base. Around eighty acres of the farm is being grazed and one of their limiting factors is being able to pipe water for livestock to some of the more inaccessible areas. Allen is excited about the recent advances in rotational grazing that make it more possible than ever to manage such a rugged landscape with livestock. Portable electric fencing, solar energizers, and advances in watering systems all help.

Being in the business of producing livestock that are slaughtered for meat is quite a stretch for the former vegetarian, but the more Allen studied ecosystems like oak savannas, the more he realized there's no such thing as grass without grazing animals, and these days, without an economic incentive to raise those herbivores, there's no practical reason to have them on the midwestern landscape.

"Yes, *Bos taurus* is not a native species, but it's a pretty close mimic," he said. "It's a lot better to have the function of herbivores on the landscape, even if they're not native, than to be insistent on only natives, because it's a little late for that."

At one point, Allen showed me specifically what happens when those herbivores are excluded from the land. We examined a five-acre piece that the farmer estimated had been in corn for around 120 years before he planted it to prairie in 2015. He had used funds from a U.S. Department of Agriculture Natural Resources Conservation Service pollinator habitat program to seed the spot with native prairie. The planting was coming along nicely, and once the contract expired, Allen planned on making it one of his livestock grazing paddocks. Why was animal disturbance an important part of this natural habitat's future? We drove up the hill that dominates the farm to see the answer firsthand.

On the lower portion of the slope, an open pasture had a few trees interspersed—it had been exposed to sheep and cattle by the previous owners over the years. Suddenly, we came to a spot where the pasture seemed to have hit a rock wall. In reality, it was a barbed-wire fence that had kept livestock out of the upper reaches of the hill to maintain "hunting habitat." The result was a dense stand of mature, 250-year-old oaks, 50-year-old maples, and other hardwoods, but also brushy undergrowth that made it all but impenetrable.

"There's not a blade of grass in there—I mean it's bare soil in there," said Allen, adding that despite the impressive timber growth, the undisturbed woodland was much less diverse than the open mix of pasture and trees below it.

Learning how to manage such a habitat hasn't been easy. The farmer conceded it has been a big adjustment to take what he had studied in academia and apply it to the land. "There were things I thought I understood from a book, but when you actually see it firsthand, it's different. I always thought of plant communities as being relatively static year-to-year and being in the same place. I had no idea how dynamic they can be."

He has adjusted his land management through observation as well as trial and error. Allen has also benefited greatly from the advice of neighboring farmers, who have insights into local weather and soil and plant conditions that a lifetime of university learning could never replicate. His long-term goal is to create a 50/50 mix of grass and tree habitat throughout the farm. Ironically, if one were to look at an aerial photo of his holdings, the conclusion would be that Allen has already accomplished this: it's pretty much evenly divided between grass and trees. But location is everything when it comes to overall ecosystem health. The problem is, these habitats are clustered together—the trees tend to be on the tops of the hill, and the open areas lower down, where they are more accessible to livestock and cropping.

"We've got 50 percent of 100 percent canopy and 50 percent of 100 percent grass. They're not integrated, they're segregated," said Allen. As a result, he's cutting down trees to open up where it's solid canopy, and planting trees in the open grassland. If that's not a recipe for removing all doubt amongst your neighbors that you've gone around the bend, then I don't know what is. But Allen's point about the segregation of habitat reminds me of what's happening on a larger scale in the Midwest. We've got ecologically healthy spots such as wildlife refuges, national parks, and forest preserves. And then we've got the places where we raise food. Satellite images would show that these two kinds of landscapes are highly segregated—they're two ships passing in the night.

Creating that interspersed habitat means some good old-fashioned grunt work involving chainsaws and brush clearing. In a way, Allen sees himself trying to replicate what the mastodons that used to roam this valley did: removing entire trees and opening up the landscape, creating a mosaic effect that attracts grazers. When one is sweating over the chainsawing of a few maples, elms, and ironwood to let in the sunlight, thinking about the profound impacts megafauna had on the landscape over millennia puts things in perspective.

"I've cleared a few acres—I've got a hundred to go," Allen said, his voice trailing off as he laughed. "I turn thirty-four next month—I have plenty of time."

Signs of Life

"Corn, corn, corn, corn, corn, beans," is how Brooke Knisley summarized the general landscape of her neighborhood while taking a break from picking beans—the green kind, not soy—on Alternative Roots Farm one evening in late summer. She wasn't exaggerating: while driving to Brooke and her husband John's vegetable farm and orchard in southern Minnesota, I negotiated a curve in the road where corn crowded in so closely I felt like I was driving through one of those car-sized holes bored through giant redwoods during a less enlightened time. It gave one a sense of monocultural myopia.

But once I arrived at the Knisleys' four-acre island of biodiversity, and took a tour of the strip of restored tallgrass prairie that separates their garden and orchard from all those acres of corn, my brain was challenged to take in all the sights, sounds, and smells—it was like plunging from a desert island into the waters of a tropical ocean reef exploding with life. First the sights: wild bergamot, grayhead cone-flower, golden alexander, blazing star, anise, black-eyed Susan, golden-rod, and, perhaps the coolest named one of them all, rattlesnake master. "I was hoping so much to get this one to grow," said John, as he grabbed the bristly edges of the rattlesnake master's leaves. "It's the only native yucca plant in Minnesota. It's cool." With equal excitement, he pointed out a cup plant, a tropical-looking sunflower-like aster whose leaves, indeed, form cups. Water collects in these natural bowls, which gold-finches often drink from. A red admiral butterfly fluttered by as I ex-amined it.

Then the sounds: the air was an electric cacophony of bees (twelve different species by the Knisleys' count), wasps, and innumerable other insects, as they droned from flower to flower, completing the sexual dance so key to flora and fauna. As the summer evening wore on, the undulating thrum of cicadas started up. I was reminded of something environmental writer Craig Childs penned after spending two days backpacking in a north-central Iowa cornfield: "I listened and heard nothing, no bird, no click of insect."[3]

And finally, the smell: I noted some bees emphatically working over the purple flowers of a mint called anise hyssop. John invited me to take a flower, crush it, and hold it up to my nose. The overwhelming but pleasant aroma of licorice engulfed me immediately.

At first glance, this thirty-foot buffer of fragrant hurly-burliness simply served as a kind of demilitarized zone between the Knisleys' plot of garden/fruit trees and a strapping stand of their neighbor's corn. But as my short hike in among the forbs and grasses that made up this prairie made clear, it was more than that—it was a significant link between Alternative Roots Farm's food production enterprise and the natural environment. And the food produced in the midst of this natural habitat, in turn, was a link between the land and the eaters who purchase the vegetables, fruit, and pork produced here.

"Yeah, it's fun to see it come to life," said John, referring to the prairie *and* the micro-farm itself.

Finding a way to bring the land back to life was the Knisleys' goal back in 2011 when they moved to the area after getting degrees in environmental-related fields and working for conservation agencies. Brooke and John actually took the same environmental economics class while they were both at Bemidji State University, in northern Minnesota. John got a degree in environmental policy and planning, which he saw as a natural follow-up to his upbringing—he grew up in New Ulm, just twelve miles from Alternative Roots Farm, and spent his youth hunting, fishing, trapping, morel mushroom and ginseng gathering, as well as splitting wood for a wood stove.

Brooke grew up in Eden Prairie, a Twin Cities suburb. She majored in environmental studies at Bemidji State and focused on ecological restoration and invasive species management. She eventually got her degree at the University of California-Santa Cruz, where, perhaps not surprisingly, John proposed to her inside of a giant redwood tree.

Both Brooke and John worked for conservation agencies in the Bemidji area after college. Brooke was with the Natural Resources Conservation Service for a time, and John was on staff with a local Soil and Water Conservation District office. Those jobs, among other things, involved working with farmers to help them to adopt basic conservation measures like wildlife plantings and grassed waterways for erosion control. Through these jobs, the couple garnered insights into what measures and government incentives are available for such practices; it also sparked an interest in getting out on the land and practicing some of this stewardship themselves.

"The entire time you're trying to incentivize people to do better on the land," said John of this government agency work. "We were kind of armchair conservationists, and wanted to put it into practice on our own. I mean what would be better than to do it ourselves, and really practicing what we're preaching and kind of living it?"

But was there a way to do it that could combine their backgrounds in environmental education and outreach with on-the-ground ecological restoration? While living in Bemidji, the Knisleys belonged to a CSA produce operation, and were impressed by the connection the owner/operators of the enterprise had with the eaters who purchased shares in the farm. What if they could use such a connection to raise food in a way that was in sync with nature, and thus help the wider community develop a closer relationship with the land?

In 2011, the Knisleys moved back to the New Ulm area when a job became available for John in a county planning and zoning office, doing environmental education and outreach, among other things. It was a great opportunity, but the reality of moving from Minnesota's lake country to what amounted to an agri-industrial landscape soon set in.

"It became *really* evident when we first came down here that like, wow, we live in the breadbasket of the United States, but no one raises anything you can eat," recalled John. "It's kind of crazy."

In fact, finding a farmstead suitable for raising vegetables on was difficult—everything was being sold off in large allotments hundreds or even thousands of acres in size. In many cases, the houses and out-buildings had been bulldozed to make room for more row crops. In fact, the first two farmsteads the couple looked at have since been plowed under. They eventually bought four acres near the small community of Madelia—the parcel had originally been part of a 160-acre farm and it consisted of a house and a few outbuildings. It was basically a blank slate, which was what the couple was looking for.

They soon set about establishing a large garden and planting fruit trees, and converted an old chicken house into a produce-packing shed. During the winter of 2011–12, they took Farm Beginnings, a beginning farmer course offered through the Land Stewardship Project. One re-quirement of that course is that participants develop a vision for their farm and set long-term goals. Like many young farmers with lots of energy, the Knisleys had big plans for their blank slate, and soon had to pare back their ambitions to fit the reality that at least one of them would be working off the farm full time. Although they may have wavered a bit on the details of just what kind of farming they were going to do, one part of their plans remained deeply rooted.

"The farm as ecosystem—we've never really seen any other way than to do it that way," said Brooke definitively. "It's always been our goal, to make it as diverse as possible and nurture those different parts—pasture and crop rotation and little pockets of wildlife—wherever we can." Just as important, she added, their goals center around nurturing people's connections to the land. "That's why we chose CSA—we want that direct connection with our customers, and for them to know directly how their food is raised, but also for them to know how it impacts our ecosystem."

The Knisleys are now raising certified organic vegetables on half-an-acre for a small CSA enterprise and a local farmers' market. Through the CSA and other share agreements with customers, the farm serves about fifty families. They recently erected a passive solar deep winter greenhouse on the farm so they can offer a winter version of the CSA. The farm markets a little over a dozen pasture-raised hogs a year di-rectly to consumers. The pigs, which graze between fruit trees and at the edges of the garden plots, create the kind of closed-loop nutrient cycle the Knisleys believe is integral to developing a healthy ecosystem.

Perhaps the fastest growing part of the farm business is the fruit enterprise, which consists of fifty varieties—mostly apples, but also plums, apricots, pears, and raspberries. The couple raise the apples on their own place, as well as a small orchard they manage near New Ulm—it had been neglected for a decade but the owner is allowing the Knisleys to manage it as long as they do it organically. They market the fruit through the CSA, as well as at a pair of food co-ops, and have some 350 apple trees on their home place that are set to begin producing in the next few years. The couple recently purchased an acre of land adjacent to the farm; they hope to plant more fruit trees on it and use a Quonset hut that's on the property to process fruit.

After growing quickly the first few years, Brooke and John, who are in their midthirties, decided to level out their size and customer base, giving them a chance to concentrate on managing the land as an ecosystem. The farm currently may not generate enough profit to sustain both of them, but the Knisleys are proud of the fact that they don't subsidize their agricultural enterprises with outside income—it's self-sustaining.

The farmers could expand their market by selling into the Twin Cities region, but serving their local community is important to them. "It's our local community that's supporting us and what we're doing and we need to have reciprocity and do the same for them and provide them with access to good food," said John.

And that means access to a feeling that by supporting Alternative Roots Farm, eaters are also supporting a more sustainable way of treating the land. After a while, "dinosaur kale is dinosaur kale," as John put it. But they hear from customers, some of them natural resource professionals, about how they like what they are doing with establishing and maintaining natural habitat in and around their food-producing areas.

Differentiating Alternative Roots kale or apples from other produce available at the farmers' market or grocery store means telling a story about how that food is the product of agro-ecological integration. "So, they know when they get that apple, it's not just an apple that came from a tree. The pigs were there, and they're helping control all the pests and all that. It's all part of a system," said John.

As he said this, we were looking at the various pig pastures bordering the edges of the farm. In one, hogs were browsing a wide variety of forages: crimson clover, white clover, red clover, and alfalfa. Swine, with their propensity to root, can be hard on a pasture. But what was striking was that the Knisley pastures that had pigs in them just a few weeks before were almost completely recovered. Lush green growth blanketed the ground between rows of fruit trees—pigs had been rotationally grazed there just that spring.

"We have a tight system," said Brooke, adding that through their newsletter and website, they extend environmental education to the food itself and the stewardship of people's own bodies through healthy eating. Off the farm, John is able to use his relationship with the land at his job in the county's planning and zoning office. He does everything from coordinate recycling and inspect septic systems, to administer wetlands regulations and conduct education programs for school children. "When I talk to kids I talk about soil and geology and I throw in pollinators and tell them it has to do with water planning," John, who is a Master Naturalist, said with a conspiratorial laugh. He's also excited to be serving on a local soil health team, which consists of a couple of dozen area row-crop farmers who are trying out soil-friendly practices such as cover cropping and minimum tillage.

Such education goes both ways: the Knisleys enjoy working with researchers so they can learn more about the impacts of their practices and ways of monitoring them. The University of Minnesota Plant Pathology Department is doing a native mycorrhizae fungi trial at the farm. Their deep winter greenhouse is also part of a state research initiative on season extension.

"Good nerdy fun," said Brooke with a laugh.

At one point, she and John showed off a research initiative that the untrained eye could easily miss. Tucked away under the eaves of outbuildings were a dozen or so rectangular blocks of wood with varioussized holes drilled into them. A close inspection showed some of the holes had bits of vegetation dangling from them. Residing inside were grass-cutting wasps, mason bees, and leaf-cutter bees. This was part of a project Alternative Roots was doing with the U of M Bee Lab, which is in the thick of trying to figure out ways of keeping both domesticated honey bees and wild pollinators like bees and wasps from disappearing completely in farm country.

It's a perfect research project for a farming operation that not only relies on pollinators for its economic livelihood but sees their presence as a key indicator of ecosystem health. All that the Knisleys do, from figuring out their gardening rotation and selecting what fruit varieties to plant, to seeding prairie and moving pig pastures, is filtered through the lens of how to make the farm a working ecosystem. Such a big picture view can be overwhelming at times without a solid pivot point to work from, one that is so basic and foundational that anything done to support it provides wide-ranging benefits throughout the entire ecosystem. When that indicator is doing well, it's a sign the rest of the system is healthy. For the Knisleys, that pivot point is pollinators.

That means keeping flowering plants on the land as long as possible throughout the year. Entomologists I've talked to repeatedly make the

John and Brooke Knisley check out a restored prairie that separates their organic vegetable plots from their neighbor's conventional row-cropped fields.

point that pollinating insects don't just need flowering plants in June and July—they need something to feed on basically from early spring until the final frost of the season. That's why Brooke and John grow a wide variety of flowering shrubs—red-osier dogwood, gray dogwood, elderberry, juneberry, serviceberry, and hazelnuts—and planted fifty different kinds of forbs when doing the prairie restoration. Sometimes, it's the small things that count: like leaving a compost pile unturned so bumblebees can nest safely, or making sure there is some sort of plant cover between rows of fruit trees. The farmers realized that pig wallows can produce micro pools of water, a valuable source of mud for pollinators that require it for their nesting material (the wallows also provide water for birds that evolved with the prairie potholes of the region).

On this particular day, John's love for pollinators was being tested—he was limping from where a wasp had stung him on the foot. "I'm really hating on wasps for a while and coming to terms with the fact that they are probably the best insect in the entire area," John said as we started a walking tour of the farm. "And I was hating on the wasps the year they tried to nest in the packing shed with me," Brooke added with a laugh.

Other invertebrates also offer indications that the land has come to life. Just a couple of years into their use of rotations, composting, and habitat establishment on ground that had formerly grown chemical-intensive row crops, they noticed what Brooke called an "explosion" of beneficial insects. One video they recorded while weeding the garden shows soil clinging to roots, and bugs, in turn, clinging to the soil.

"When we took the weeds out of the wheelbarrow, all the soil's just crawling with life in there, whether it's ground nesting beetles, baby lady bugs, ants, or what have you," John recalled. "I know we have a lot of predatory beetles in our garden, because we do a lot of mulching with straw, so they have places to hide and hang out."

But all that life basically stops at the property line. Alternative Roots is a garden spot in an industrial park. Despite the contrasting manners in which they manage the land, the Knisleys say there have not been any major clashes with their monocropping neighbors, who tend to be conscious of not spraying on days when prevailing winds may send toxins into the garden at Alternative Roots, for example. Such neighborliness goes both ways: the Knisleys work to keep what row crop farmers would consider noxious weeds from spreading over the property line.

"We have this island with all these really great flowers and insects and everything, but unfortunately this is the only place they have to live," said John.

As he said this, I looked past the pig pasture. A solid mile of corn and soybeans separated Alternative Roots from a low-slung confinement hog facility. I grew up around swine production, but later, as I drove past the hog barn, the plume of odor escaping the operation slapped me like a scoop shovel to the face. Gates closed off the driveway, and a "No Entry" sign was posted.

There was no sign of life.

The Curse of Knowledge

A few years ago, when Bryan Simon returned for a visit to his alma mater, the University of Minnesota–Morris, he ran into one of the professors he had studied under while getting a bachelor's degree in biology at the school.

"He asked me what I was doing, and I said, 'Oh, I'm farming.' And you could just see the look on his face," Simon recalled. "It's like, 'Oh, what a waste—you have a liberal arts degree in biology and now you're farming.' And I was like, 'But you've got to understand what *kind* of farming I'm doing.'"

As Simon related this story, he and his wife Jessie were sitting at the kitchen table of a former hunting lodge on 195 acres of farmland in west-central Minnesota's Grant County while their two children, Charlie, four, and Annella, two, played in the yard on a wet July morning. Beyond the yard was Lakeside Prairie Farm, an enterprise characterized by a mix of restored prairie, rotationally grazed livestock pastures, and oak savanna habitat, broken up here and there by small wetlands. The farm thrusts peninsula-like out into nine-hundred-acre Cormorant Lake, where ducks, along with Canada geese and pelicans, could be seen floating on the water. It was clear the *kind* of farming the Simons were undertaking blended the principles of ecological restoration and wildlife biology with good old-fashioned agronomy and animal husbandry.

But as is evident by the view beyond the Simons' driveway, a different kind of agriculture dominates the majority of the midwestern landscape: acre after acre of annual row crops like corn and soybeans, a duo-culture covering soil just a few months out of the year on land once dominated by prairies rich in hundreds of different species. And as evidence mounts that everything from grassland songbirds and waterfowl to pollinator insects and amphibians, not to mention water quality, is suffering as a result of loss of habitat at the hands of industrialized row crop agriculture, conventional farming is not exactly viewed as friendly to long-term environmental health.

Given all that, it's not surprising that a professor dedicated to teaching about natural resources protection would be disappointed to learn a former student had gone into farming. Perhaps it's akin to an art major becoming a highway engineer.

In some ways, the Simons' passion for nature was sparked about as far away from the farms and small towns of west-central Minnesota as one can get. A biology teacher at a high school they attended in St. Cloud would regularly lead teens on month-long canoe trips into the Canadian wilderness. Both Bryan and Jessie, who are now in their midthirties, participated in these trips, which went as far north as Hudson Bay (they were a few years apart in high school, and didn't meet until after graduation).

"That made me want to get into conservation," recalled Bryan of those trips. "Being in a pristine ecosystem with no visible human impacts—to be able to observe that and live that for a month was eye-opening."

The Simons carried that passion through high school and into college. Jessie ended up getting a master's degree in environmental education from Hamline University and now teaches second grade, where she uses her environmental background as much as possible in the curriculum. "I try to do things with my students that get across the message that your actions affect more than just you," she said. "I try to be intentional about going out as a class and taking note of phenology throughout the year."

After getting his biology degree at U of M–Morris, Bryan did seasonal work with the Student Conservation Association, which placed him with the U.S. Park Service and the Bureau of Land Management in places like Idaho, Texas, and Hawaii. Through that work, he led crews of interns, doing invasive species control and native seed collection. Bryan later worked for The Nature Conservancy in eastern South Dakota, where he did fire management and plant monitoring.

The ideal natural resource career trajectory, right? But nestled back in Bryan's mind was a seed of an idea about the role working farmland conservation could play in restoring and maintaining habitat. Yes, he had seen pristine wilderness untrammeled by humans, and yes, he had worked for organizations that protected natural areas. But while growing up in St. Cloud, Bryan had frequently visited his grandparents' farm near Morris. There he realized that the environmental fate of the majority of the midwestern landscape is in the hands of farmers, who are out there working the fields daily. One day, there was a discussion going on in a college landscape ecology class about whether consumers or farmers have more responsibility for the way food is produced and its environmental impacts.

"I took the position that the farmer could have the greatest influence on the landscape," recalled Bryan. "They have the most control over land use and they ultimately decide how well the land is taken care of."

Later, while pursuing a master's degree in ecology at South Dakota State University, Bryan conducted research at EcoSun Prairie Farms near Brookings, which had been set up by one of his professors as a large scale experiment to determine if grass-based farming could make returning prairie to the landscape a profitable venture.[4] Bryan, by this time passionate about prairies, was inspired by the experience. He became convinced there was a way to make it so natural habitat and significant food production could occupy the same piece of ground.

While in graduate school, Bryan took Farm Beginnings, enrolling in the class with Ryan Heinen, a friend of his since seventh grade who had a similar academic/professional background in natural resources. Simon and Heinen share a passion for the prairie, and thus went into the Farm Beginnings class knowing what kind of farming they were going to do: grass-based livestock production. Innovations in managed rotational grazing systems, portable fencing, and pasture improvement in recent decades have made it possible to graze cattle and other livestock on grasslands in ways that not only improve forage quality and extend the grazing season, but benefit habitat for wildlife like grassland songbirds and pollinators. In recent years, managers of nature preserves and wildlife refuges have recognized the benefits of utilizing rotational grazing as a way to control invasive species in prairie systems and maintain healthy grassland habitat (see chapter 3). Graziers, for their part, like that the native warm-season grasses and forbs in prairie systems can help get them through the traditional "summer slump," when the cool-season grasses found in domesticated pastures tend to go dormant.

As a requirement of the Farm Beginnings course, Simon and Heinen had to develop a business plan, which turned out to be a critical tool for getting access to the 195-acre farm in Grant County. At a sustainable farming conference they were introduced to Joe and Sylvia Luetmer, Alexandria, Minnesota, residents who were looking to buy a farm and get some beginners started on it. The Luetmers liked the young farmers' plans for utilizing rotational grazing and other methods to support a healthy farm landscape. Soon after, the Grant County farm came up for sale. The owner had been renting out the tillable acres for corn, soybean, and wheat production, along with utilizing the small house as a hunting cabin. There are approximately twenty-five acres of wetlands, and a remnant of oak savanna dominates one end of the property. In other words, it was perfect for what Simon and Heinen had in mind: start a farm that blended the wild and the tame. In 2012, the Luetmers bought the farm and began renting it to the beginning farmers. The Simons moved into the former hunting cabin, and Bryan and Ryan started putting in place their eco-based farming operation by clearing out invasives, converting cropland to prairie and erecting fencing for rotational grazing of livestock. In 2016, Ryan and his wife Barbara decided to pursue their own dream of grass-based dairy farming and moved to a different part of the state.

The Luetmers have agreed to eventually sell the property to Bryan and Jessie at the same price they originally bought it for. In addition, the Simons are utilizing a U.S. Fish and Wildlife Service program that puts the farm's grassland acres in a permanent easement. The easement allows grazing of the acres, as long as they are kept in perennial grasses. The ultimate effect of the arrangement is that it reduces the "economic value" of the farm, since it can't be cropped or otherwise developed. That, along with wetland banking credits they hope to get from the government, will eventually make the land much more affordable to Bryan and Jessie.

"That's the best thing we've got going for us right now, as far as the economic side of farming," Bryan told me, only half-joking, adding that the perpetual nature of the grassland easement is both good and bad. "The good is all of our work here restoring the prairie and putting in all this high diversity mix will be preserved. It will be prairie, hopefully, forever. But from a farming point of view, who knows what the future holds? It does tie the hands of the future generation a little."

On the summer day I visited the farm, the future generation was enjoying the open landscape of the here and now. Charlie and Annella accompanied their parents as they walked up the farm's long driveway to check on a twenty-acre piece of land that the year before had been converted from row crops to an eighty-species prairie planting. Perimeter fencing had been erected so the prairie could be grazed; on the other side of the fence in another former crop field, a recent seven-acre planting of rye and oats was preparing the soil for the next grassland seeding. The restored prairie was doing well: the yellow of brown-eyed Susan plants added a bright pop to a hillside shrouded in a July mist. Prairie phlox was also thriving, as well as, to the Simons' chagrin, plenty of Canada thistle. Bryan wasn't happy about the fact that they had to use herbicides to control the thistle in order to get the prairie established, or that in actuality a prairie like this should have three hundred different species represented. But such compromises are the bargains one must strike when undertaking ecological restoration in farm country.

An eastern kingbird and a dickcissel called out from pastureland across the driveway. The Simons have identified ninety-nine different species of birds on their farm. "Well, now, we've only *identified* ninety-nine different species," said Bryan sheepishly. "There's more here that I haven't put a name to yet."

Bryan and Jessie Simon are converting former row crop acres to prairie that they are grazing.

Jessie, Bryan, Charlie, and Annella Simon stand in a restored prairie on their west-central Minnesota farm.

In the hilly pasture, twenty-nine head of beef cattle, representing various breeds—British White, Angus, Devon, and Hereford—grazed. Closer to the house, two sows served as the foundation of the Simons' new pastured pork enterprise. Beyond the pig pasture, there were glimpses of Cormorant Lake through the understory of a stand of 150- to 200-year-old bur and white oaks. This view of the lake was a result of a labor-intensive buckthorn removal effort that's ongoing; over seventeen acres of the invasive had been cut at the time of my visit, some of which was being burned in the Simons' wood stove as a kind of red-hot revenge.

Bryan and Jessie are hoping a combination of mechanical invasive species removal and utilizing livestock grazing to keep the understory open will bring back forty acres of oak savanna habitat on this farm. Sedges, jack-in-the-pulpit, Dutchman's breeches, and snow trillium were already responding to the opening up of this habitat—as was, unfortunately, the invasive weed burdock.

They were also using livestock grazing to thin out the reed canary grass and cattails that were choking out the shallow marsh and wet meadow regions of the farm's wetland habitat. This runs counter to a common misconception that wetlands and livestock should never mix—just as native grassland restoration can benefit from animal disturbance, so too can semi-aquatic habitat.[5]

The Simons are aware the livestock aren't just here to maintain natural habitat—they, and these acres, must earn their way. Lakeside Prairie Farm started out also producing vegetables, chickens, eggs, oats, and wheat, but they eventually narrowed its enterprise focus to grass-fed beef and pastured pork. When I visited the farm, Bryan and Jessie were direct marketing about a dozen head of cattle and approximately the same number of hogs each year, which wasn't enough to make the farm financially self-sustaining. Their goal is to double the number of beef animals they sell, and market as many as one hundred pigs annually.

But access to grazing land limits their production potential. It takes time to refashion row-cropped acres as productive grassland, especially when one's goal is to have native species be a major part of the mix. On one part of the farm, Bryan pointed out seventy acres of land that had been idled for several years under the Conservation Reserve Program. Invasive red cedar had actually been planted on the idled ground, all but ruining it as a grassland. The contract was soon expiring, and it was clear the Simons couldn't wait to use chainsaws and cattle to bring back the grassland habitat on these acres—it would provide much-needed feed, while allowing them to test yet again the theory that farming and natural habitat restoration can be integrated.

The young farmers know that in order to attain their dream of balancing ecological health with financially viable farming, they will need help via public policy as well as the marketplace. Bryan is frustrated that the bulk of federal farm policy doesn't see diverse, ecologically healthy operations like theirs as a public good, and that it instead promotes monocultural crop production. However, they have received funding through the U.S. Department of Agriculture and the U.S. Fish and Wildlife Service to set up rotational grazing systems, seed native species, and take on invasives removal.

Marketing a product that is good for the environment can also be frustrating at a time when consumers seem to favor convenience and price over sustainability, no matter what the long-term costs to the landscape and communities might be. Bryan is hopeful consumers can adjust their priorities. After all, he himself was able to go against the conventional wisdom that a healthy environment and farming are mutually exclusive. "Once you gain that knowledge that it's not either one or the other, you're cursed; you can't go back to being ignorant," he said as we headed back to the farmyard. "And with that knowledge, you seek to bridge those worlds."

Bridging those worlds means inoculating consumers with the idea that what they eat influences not only their own health, but the health of the land. There are signs that people are willing and able to make such connections. Bryan was surprised when a customer all the way from the Twin Cities found them on the internet and ordered beef from Lakeside Prairie Farm. "I asked him why he was buying from us," recalled the farmer. "He said, 'I like what you guys are doing ecologically.'"

10

Hubs of Hope

The Connection between Inebriated Grasshoppers and Your Dinner Plate

The danger of telling the stories of innovative farmers such as those highlighted in this book is that they can be seen as too much of an anomaly to be replicated. When Gabe Brown says, "There are people all over doing this. They just don't have the mouth I have," what he's trying to convey is that his only outstanding attribute is his willingness to go public with his hits and misses. Indeed, for every Gabe Brown who's on the speaking circuit, hosting international visitors, or starring in online videos, there are dozens of ecological agrarians who are more quietly blending the wild and the tame.

But there's not enough of those kinds of farmers. The bulk of U.S. agriculture is as far removed from natural processes as a factory making circuit boards. And we're paying the price in terms of dirtier water, sickened soil, out-of-control greenhouse gas emissions, decimated wildlife populations, and shuttered Main Streets. I will admit to a bias here: I believe there need to be more farmers on the land, not fewer. After spending so much time on agricultural operations of all kinds, I'm more convinced of that belief than ever. Wildly successful farming requires more eyes (and ears) observing, reacting, adjusting—and that monitoring needs to take place over a lengthy period of time. The short-term decision-making that characterizes industrial agriculture just doesn't leave much room for natural processes. What works one year may not work again for several years down the road—if ever.

One huge advantage wildly successful farmers have over their more conventional brethren is a willingness to share information. That may sound strange, given rural America's reputation for working as a collective; for example, consider the farmers' co-op movement that revolutionized the grain trade.[1] As I've written about in chapter 6, it was the willingness of innovators and early adopters to share their experiences with their neighbors that led to the rapid spread of hybrid seed corn.

But in the early 1990s, when I was working for a mainstream farm magazine that had as its readership some of the largest farmers in the country, I began running into a troubling trend. Some of these farmers were unwilling to be interviewed for stories about a particular innovative production or marketing technique they were using. "What's in it for me?" was a version of the response I would get over the telephone. They expressed concern that sharing their "trade secrets" would put them at a competitive disadvantage with their neighbors—who they now saw as rivals—for land, market share, and profits. For someone who grew up in an era when farmers still got together to shell corn or bale hay communally, this was a real eye opener.

An increasing number of farmers were raising an increasingly undifferentiated product: corn and soybeans for the international grain trade, for example. When one was in a position to take advantage of a market opportunity that paid a little bit more—high-oil soybeans or extra-lean hogs, for example—the last thing they wanted was other farmers horning in on their financial success. And who could blame them? The agricultural economic crash of the 1980s put a large number of farmers off the land in a very short time. The path to profits became paved with raising more bushels on more acres (or more pounds of meat and milk per square foot of barn space). To survive, you had to be a hard-nosed business owner willing to expand constantly, often at your neighbor's expense. The trouble is, that get-big-or-get-out attitude didn't do those tough-talking farmers any good either. There was always someone bigger and more powerful to take market share. And that competitor wasn't necessarily in the next township or county—the Cargills of the world don't care if they buy their soybeans in Iowa or Brazil, as long as the commodity is as cheap as possible. Thus, farming has become what University of Missouri economist John Ikerd calls "a race to the bottom"—a race fueled by exploitation of land and people.[2]

On the bright side, I've witnessed in the past decade or so somewhat of a return of the farmers-helping-farmers culture. Actually, I suspect it never completely left, but just got overshadowed by the economic storms raging over rural America. Every week I run across examples of farmers sharing information and ideas openly. Partly it's because when someone is doing something truly innovative—utilizing cover crops to cut fertilizer use and suppress weeds or using mob grazing to double a pasture's ability to produce livestock, for example—they're excited about it. It's human nature to share such breakthroughs and get feedback on how to make them even better. I'm seeing even "conventional" farmers more willing to share with their neighbors these days. That's particularly true when it comes to the current revolution in building soil health. As the corn and soybean farmers in Indiana are learning, injecting just a bit of the "wild" into their otherwise domesticated row crop fields can produce tremendously positive results. That's exciting, and fun to share.

The internet and social media have made the trading of this information simpler than ever, creating communities across thousands of miles. Log into any e-mail listserv where people share innovative ideas about farming closer to nature, and you're likely to feel pretty positive about the future of agriculture.

It's not just the thrill of discovery that motivates farmers to swap ideas. Many agrarians I've interviewed in recent years are also adherents to the philosophy of writer/farmer Wendell Berry, who would rather have a neighbor than have his neighbor's land.[3] Dan Jenniges sees a direct connection between more grass on the land and more beginning farmers in his community. Marge and Jack Warthesen host beginning farmer trainings and have mentored newbies. Having the most wildly successful farm in the county means little if the rest of the community is basically abandoned. As writer Michael Pollan puts it when describing the "remade" state of Iowa: "The only thing missing from the man-made landscape is . . . man."[4]

Perhaps the most positive trend I've witnessed in recent years is how beginners with little background in farming (or rural living) have been welcomed into agricultural communities by lifelong residents. Southwestern Wisconsin farmer Peter Allen expresses genuine surprise at how much he and his family are supported by their neighbors, even though he's a refugee from the big city of Madison, Wisconsin, and that up until the time he stepped onto those hilly acres in the Kickapoo Valley, he had spent the majority of his adult life as an academic. "I think they're just happy to see some young person out trying to farm, because none of their kids are doing it," he told me. He's right: the latest U.S. Census of Agriculture shows the age of the average farmer is fifty-eight years old, up from fifty-one in 1982. Of principal farm operators, only 6 percent are under thirty-five.[5]

Allen's warm reception isn't unusual. I've spent a lot of time in rural communities and talked to older farmers who are extremely happy to see young, energetic people participating in a kind of reverse brain drain. Sure, they may have some weird ideas about "organics" and "sustainability," but they also share with those older farmers a love of the land. Even better, no matter what kind of farming these greenhorns are undertaking, they require information on local soils, climate, and sources of inputs—details they can't glean from a textbook or YouTube. It's a basic instinct to be needed, and the generational knowledge these veteran farmers have is needed now, more than ever.

But we still don't have enough wildly successful farmers to support our communities—both natural and human. What will it take to make this kind of relationship with the land more the norm, and not a quirky abnormality? Well, first of all, I'm not sure this kind of agriculture will *ever* become commonplace, particularly when talking about an operation like the one Martin and Loretta Jaus have, which has seamlessly blended the agronomic and the ecological. In a way, their operation is a perfect storm: both farmers have academic/professional experience in natural resource management; they have livestock, which requires a diversity of plants to thrive; and they are being rewarded in the marketplace with an organic price premium. To top it off, they are very comfortable associating with the environmental community, which provides them the kind of positive reinforcement they may not get from their conventional farming neighbors.

But I do think that wildly successful farming can seep into the cracks and crevices of conventional farming. We've certainly seen this with the soil health revolution, and how methods such as cover cropping are helping mainstream corn and soybean farmers deal with such issues as herbicide-resistant weeds, compaction, and erosion. Dan DeSutter would probably still be using a deep ripper, and who knows how much diesel fuel, to break up compacted soil on his Indiana farm if he had not seen those ryegrass roots "bio-drilling" through the hardpan.

To be honest, an injection of a little bit of the wild into domesticated agriculture isn't some sort of unhindered flow of ideas and techniques. Significant barriers remain. For example, an examination of conventional farmers' current interest in soil health unearths one of the biggest challenges midwestern agriculture faces if it is to adopt on a wide-scale basis practices that contribute to a healthier ecosystem: integrating livestock back onto the land. Livestock help cycle nutrients and provide a way to add economic value to grass, small grains, and other plants that benefit the soil. "Livestock are the rock stars of building soil health," Justin Morris, a soil expert with the Natural Resources Conservation Service, told me one day while standing near a rotationally grazed pasture in southeastern Minnesota. "Nothing else quite compares to four hooves and manure."

But let's face it, there are a lot of farmers who simply do not have the capacity or desire to "go back" to having a diversified mix of enterprises that integrate crops and livestock. That's why it's important to take seriously ideas for creating "community conservation" in a region, such as what is being proposed in Dan Jenniges's part of west-central Minnesota. Such an initiative could allow farmers of all types—livestock, crop, vegetable, specialty, a mix of various enterprises—to operate while sharing resources. For example, a farmer looking to add economic value to a corner of the farm that grows good grass but not much of a corn crop, could borrow her neighbor's cattle, much like wildlife refuges use livestock from the community to control invasive species. I've already seen examples of farmers who raise cover crops in their corn-soybean systems renting out those acres to cattle producers who get a month or so of grazing in the fall before winter snows blow in. This is being done on a limited basis, but there's potential for it to be done on a bigger, region-wide scale. After all, the impacts of healthy ecosystems should be pretty much borderless.

The positive repercussions of this kind of community ecological restoration could extend the benefits of wildly successful farms beyond watershed, county, or even state boundaries. One of the more fascinating papers I've read in recent years was a research editorial published in the *Journal of Soil and Water Conservation*. The paper, which is based on an analysis of studies from around the world, describes how it's becoming increasingly clear that healthy soil has a huge potential to sequester greenhouse gases—the higher the organic-matter content of soil, the more carbon stored. For example, our soil holds three times the amount of carbon dioxide currently in the atmosphere, and 240 times the amount of gases emitted by fossil fuels annually. Increasing the amount of carbon stored in soil by just a few percent would produce massive positive benefits. And since farmers deal directly with the land, they could play a significant role in developing what's being called "climate-smart soils."[6]

As a result, any farming practice that can prevent soil from blowing or washing away, as well as keeping it biologically healthy, is going to have a major positive impact on our climate. That's why the authors of the *Soil and Water Conservation* editorial recommend a farming system that gets as much land as possible blanketed in continuous living cover 365 days a year. Their solution? Get livestock out on the land. The key phrase here is, "out on the land." Producing pork, beef, and milk in intensive confinement, where feedstuffs are trucked in and liquid manure becomes a waste product that must be stored in massive quantities before eventually getting disposed of, is a major source of greenhouse gas emissions.[7] In addition, such systems are reliant on monocultural production of corn, soybeans, and other crops. This results in emissions as a result of tillage, as well as the production of the petroleum-based fertilizers, fuels, and pesticides involved in crop production.[8]

But when livestock are raised on grasslands and other forages, the soil can be a sink for greenhouse gases, both because it is not being eroded and exposed to the elements, and because the world beneath the surface is building up soil organic carbon. It's important to keep in mind that it matters how those animals are being grazed. Simply turning them out onto open pastures or rangelands and allowing them to roam at will creates its own problems. Overgrazing destroys plant communities and is a major source of erosion and compaction, not to mention water pollution.

Rather, managed rotational grazing systems like what the various farmers I've described in this book have in place helps keep the grassland healthy above and below the surface by spreading nutrients sustainably and allowing plant life to rest and recover. The *Journal of Soil and Water Conservation* editorial cites studies showing how this system—they call it "regenerative adaptive multipaddock conservation grazing" (there's a mouthful)—can actually sequester more greenhouse gases than are being emitted. Such a system can even make up for the greenhouse gases ruminants produce in the form of belched-up methane.

What's particularly exciting about this journal article is the emphasis the authors place on integrating livestock, pastures, and crop production—a perfect mix of enterprises in the Midwest, a mix that I've seen firsthand on many wildly successful operations. They outline a working lands scenario where a carbon-trapping farm may have some permanent pasture that is broken up into rotational grazing paddocks.

But it could also be producing corn and soybeans in a system where cover crops like cereal rye or tillage radish are used to blanket that row-cropped land with growing plants before and after the regular growing season. These cover crops could provide low-cost forage for cattle and other livestock, taking pressure off the perennial pastures and helping justify the cost of the cover crop establishment while protecting the soil from erosion and building its biology. Cover crops can also cut a farm's reliance on greenhouse-gas-producing chemical fertilizers.[9]

The paper outlines the greenhouse gas emissions potential of several farming scenarios in North America: from keeping our current industrialized system (an increasing amount of grassland plowed under to make way for row crops while keeping livestock confined in large concentrated animal feeding operations) to utilizing a combination of managed rotational grazing and conservation cropping systems that involve no-till, diverse rotations, and cover crops. Not surprisingly, the current industrialized system would make our climate change problem worse, according to the researchers. But the scenario that involves getting more animals out on the land as part of an integrated system would build soil health to the point where agriculture becomes a net carbon sink.

This last scenario wouldn't necessarily require every farm to become a diversified crop/livestock operation. In his book *Grass, Soil, Hope*, conservationist Courtney White describes how some Western cattle producers have taken up "carbon ranching" through a combination of rotational grazing and soil improvement methods such as composting. He's describing practitioners who are committed to the business, and lifestyle, of raising livestock through thick and thin, so managing animals as carbon builders dovetails nicely with the way they do things.[10] But as I've pointed out, many midwestern corn-soybean farmers are committed to raising crops and nothing else. Under a more community-wide, integrated system, diversity could be adopted on a regional basis. Even crop farmers who do not have livestock could utilize their neighbor's animals to add economic value to cover crops or that piece of pasture that hasn't fallen under the plow yet.

Sounds great, doesn't it? But the *Journal of Soil and Water Conservation* editorial—and other scientific analyses like it—recognize that there are major barriers to integrating livestock grazing/row cropping in a soil-friendly manner, not the least of which is government policy that promotes the production of a handful of commodity crops and penalizes diversity. "Rather than reducing ruminants and encouraging destructive agricultural land use by providing price subsidies and other subsidies, rewarding regenerative agricultural practices that focus on increasing soil [carbon] and that lead to greater adoption by land managers is essential to creating a robust, resilient, and regenerative global food production system," conclude the authors of the editorial.[11]

Or, as Martin Jaus put it to me more succinctly: "A lot of the crops you see here aren't even considered crops, according to the Farm Bill. It seems that the more the crop is harmful to the environment, the more they want to support it."

Children of the Corn

He's not that far off. Time for a little Federal Ag Policy History Lesson 101. Modern agriculture policy was launched in the 1930s as a helping hand for farmers wracked by the double catastrophe of the Great Depression and the Dust Bowl. The chief architect was Henry A. Wallace, a seed corn pioneer, Depression-era agriculture secretary, and southwestern Iowa native—he was born one county over from my home farm—who sought financial support for producing certain crops such as corn. He saw government support for raising commodities as a way to ensure an "ever-normal granary." Since then, federal farm policy has evolved into an amalgam of subsidies, crop insurance initiatives, cost-share funds, and loan guarantees.[12] Despite the many reiterations it's undergone, at its core, farm policy has stayed true to what Wallace wanted: producing lots of a few crops and *lots* of one in particular: corn. "In the Midwest we are children of the corn," University of Minnesota economics professor Steven Taff told me.

Even supposed "market-based" incentives like creating demand for corn by distilling it into ethanol has a basis in policy: the industry would be a fraction of its current size if it wasn't for such federal programs as the Renewable Fuel Standard.[13] When the focus of policy is to produce commodities full-bore, it doesn't leave much room for biodiversity, or even basic measures that inject some ecological health into the system.

"The system for conservation is backward," said southern Minnesota farmer John Knisley, who, as I explain in chapter 9, has created an island of biodiversity in a sea of corn and soybeans. "We took our own initiative to do this, and there's no support for me taking land out of production at all, or planting a prairie. But if I destroy it, they'll give me money to restore it. At the same time, I feel the environment we've created here gives the land better balance, so we get a droughty year, or we get a really wet year, the land, because of the way it's been treated, can take that, whatever the environment kind of throws at it."

Indeed, I visited his and Brooke's farm during a particularly wet July in that part of Minnesota. The Knisleys' low-lying farm was thriving, a stark contrast to many of the drowned-out spots I saw in area corn and soybean fields. At one point, I saw a field where tall corn stalks became shorter and shorter as they marched toward a small pond that had emerged as a result of heavy rains. This is a reminder that this area was traditionally the home of wetlands—large marshes as well as smaller ephemeral potholes. When it rains enough, the land reminds farmers and others that wetlands aren't always an historical artifact. The land remembers what people and government policy forget.

The federal Farm Bill comes up for renewal every five years or so. Since I started my career as an agricultural journalist, I've seen five Farm Bill "cycles" (at this writing, I'm beginning my sixth). The sustainable agriculture community has been successful each round in making federal farm policy slightly more "green." For example, the Conservation Stewardship Program, which was the brainchild of the late Minnesota farmer Dave Serfling and other members of the Land Stewardship Project's Federal Policy Committee, actually pays farmers to produce positive environmental benefits—everything from cleaner water and healthier soil to pollinator and wildlife habitat restoration. Other Farm Bill initiatives, like the Environmental Quality Incentives Program, provide cost-share funds so farmers can establish sustainable infrastructure such as rotational grazing systems.

Such greening of farm policy is a positive trend, but the vast majority of the Farm Bill's resources continue to go toward growing more corn and soybeans in a monocultural, industrialized system. It's like conservation is running a race that gave industrialized agriculture a several-decade head start. The *Minneapolis Star Tribune* newspaper reported in 2015 that for every dollar taxpayers spend protecting water from the negative results of industrialized farming practices, farmers receive five dollars in subsidies that encourage them to do more of the same.[14]

At times, one has to admire the ability of the supporters of such a system—commodity groups, multinational grain traders, meatpackers, food manufacturers, and ethanol producers, to name a few—to refashion a set of policies just enough that they look like reform, but in the end, accomplish the same task: produce more and more commodity crops on more and more acres. For example, in recent Farm Bills the federally subsidized crop insurance program, which was originally set up in the

1930s as a safety net so farms wouldn't be wiped out by weather disasters, has mutated into a program that encourages crop production on land that would otherwise be considered too erosive, wet, or otherwise "marginal" to produce a decent crop. In economics, this is referred to as a "moral hazard"—basically a situation where one person takes more risks because someone else bears the cost of such risk taking.[15] Since taxpayers foot the bill for at least 60 percent of the cost of a farmer's typical crop insurance premium, we're all caught up in a bit of a codependent situation here.

A U.S. Government Accounting Office (GAO) analysis blamed crop insurance, along with the National Flood Insurance Program, for inflating the cost of recovering from disasters by increasing risky behavior. "While federal law prohibits crop insurance from covering losses due to a farmers' failure to follow good farming practices . . . some of these practices maintain short-term production but may inadvertently increase the vulnerability of agriculture to climate change through increased erosion and inefficient water use," concluded the GAO.[16]

In other words, by taking the risk out of planting row crops on land that normally would be considered too marginal to produce a profitable yield, crop insurance is subsidizing farming practices that are making our land less resilient. Providing a safety net is good policy; stifling innovation is not.

"Crop insurance subsidies are probably the biggest limiting factor when it comes to adopting innovative soil health measures like cover cropping. They incentivize practices that go against building soil health," Indiana farmer Dan DeSutter told me. "Without crop insurance, you could make the land much more diverse."

Crop insurance makes it so all those sloughs, hilly pastures, and just plain odd corners on a farm that are difficult to raise a good crop on—think the Jack and Marge Warthesen farm—are now worth tilling. And as cattle producers have discovered, when you are renting pasture ground that was traditionally considered too rough to be row cropped, crop insurance can suddenly make it lucrative for your landlord to plow up the grass, tear out the fence, and plant corn and soybeans. And that's exactly what's happening. Analyses of land use patterns have made a connection between the growth of crop insurance and the destruction of grasslands, wetlands, and other marginal habitats.[17] Farming is inherently risky, given the vagaries of weather and markets, and that's part of the reason programs like crop insurance were created. But there's a difference between cushioning the blow and fueling endeavors that have negative consequences. In many ways, the morphing of crop insurance from a basic safety net into a risk promoter that damages the landscape is yet one more example of a regrettable substitution in agriculture.

Public Good

The crop insurance example illustrates that until wildly successful farming is recognized as a public good, the overall farm policy bias toward monocultures will remain, plain and simple. The stakeholders that drive agriculture policy—major agribusiness firms and commodity groups—simply make too much money from the current dysfunctional mess. Real change will need to be triggered by a constituency that is generally silent during farm policy discussions: the general public. Specifically, people who care about clean water, wildlife habitat, the climate, and healthy soil need to make a connection between how the land is farmed, what they eat, the state of the environment, and what policies they support.

We now have a farm policy system that consistently forces many farmers to undertake activities that go against their own ethical standards. Yale University psychologist Stanley Milgram is famous for his experiments in which study subjects thought they were advancing science by exposing humans to painful shocks. "Ordinary people, simply doing their jobs, and without any particular hostility on their part, can become agents in a terrible destructive process," Milgram once said.[18] That could describe a lot of farmers.

For farmers like the Jauses, ag policy's bias against the way they farm is not about the money, although they do lose out on certain kinds of subsidies by being so diverse. Perhaps an even bigger blow is that federal farm policy, by ignoring or penalizing Loretta and Martin, is sending a strong negative message: your kind of farming is not valued by society, it is not a public good.

No, not all farms in the United States will become wildly successful. But we do need enough of these kinds of operations to serve as models of what to strive for. I see them as nurturers of innovation that other farms can selectively harvest ideas from. A wildly successful farmer's methods for improving soil health could be adopted by a corn-soybean farmer. Rotational grazing systems could be copied by a conventional dairy or beef producer. Establishment of pollinator habitat benefits a vegetable or fruit producer who needs the services those insects provide. A corn farmer could "borrow" the livestock of a diversified neighbor to add economic and agronomic value to a cover crop. If such partnerships take off, there may be a time when wildly successful farms are seen as a positive addition to a rural community, and not just because people enjoy seeing bobolinks or want to hunt pheasants. Such farms could be considered integral to the long-term sustainability of row crop agriculture, especially as the downside to a chemical- and energy-intensive, tillage-centric system becomes increasingly evident. Who knows? Commodity groups like the National Corn Growers Association or American Soybean Association may someday use their immense lobbying power to promote policy and public research that supports wildly successful farming.

In the meantime, a handful of wildly successful farms cannot be the hub of inspiration for the entire country, or even a state or county. We need more of them to inspire change in every watershed or township. How many? When asking that question, I like the answer Tom Will, a bird ecologist for the U.S. Fish and Wildlife Service, gave during a recent visit to the Jaus farm: "I don't know how many we need, but I know how many I want—when I look at the landscape and it feels right."

In other words, we should strive for as many as possible. Get enough of them, and we don't just have wildly successful farms, but wildly successful communities, which is what big picture ecosystem health is all about. That brings us back to creating the right atmosphere for developing as many of the next generation of wildly successful farmers as possible. There are plenty of young people interested in blending the wild and the tame. But to encourage them, and inspire others while supporting them financially, we need to create that constituency I described earlier. Such a constituency can be generated around a big picture issue, such as clean water, or a smaller issue, like protecting grassland songbird habitat. I've even seen people who are concerned about the future of red-headed woodpeckers reach out to farmers like the Warthesens who are willing to keep a dead snag or two in a fence line.

There is also a constituency that includes all of us: eaters. In his book, *After Nature: A Politics for the Anthropocene*, Jedediah Purdy makes the case that the relatively new foodie movement—people wanting to know how and by whom their food is produced, and putting their money where their mouth is—can help make a critical connection between how farmland is treated and what's on their supper table. Purdy argues that for too long the environmental movement saw "ecological farm-ing" as an oxymoron. Before that, nature was not really nature if it was trammeled by humans.[19]

Big picture environmental change came about in the 1970s and 1980s when the dumping of raw sewage and toxins was seen as a public health threat. Here in the U.S., it's rare to see a factory or a municipal-ity sending their waste straight into the water anymore. Stopping such "point pollution" practices didn't require the general public to identify with factory owners or even a city council member. It just required raising enough hell to trigger regulation of some of the most outrageous practices.

But when it comes to agriculture, things get trickier. All those thou-sands of farms of various shapes, sizes, and structures don't lend them-selves to cookie-cutter environmental regulations. The largest concen-trated animal feeding operations can be regulated in terms of how they dispose of the millions of gallons of liquid manure they produce, or how much hydrogen sulfide they emit, but what about the family-sized opera-tions raising a few dozen pigs on pasture? In that case, outright regula-tion may not only be ineffective, but it may actually stifle innovation. It may also put that smaller farmer out of business, and that land will be gobbled up by the larger CAFO.

Purdy says that true environmental change requires people to iden-tify personally with the way business is being done. People's interest in how their food is produced offers such an opportunity. Farming could become a public good, much like green space or a national park. "As a cultural matter, the food movement offers a way to make abstract eco-logical values concretely one's own," writes Purdy.[20]

Part of the reason environmentalists have long wanted to separate nature from working farms is the ongoing threat our pristine wilderness areas face from development—everything from mining, drilling, and recreation to, of course, agriculture. Blurring the lines between food production and working ecosystems could be just one more way to justify encroaching on areas like Yosemite, Zion, or Bob Marshall. That's a justifiable concern. There are certain natural jewels that should be off limits to even the most ecologically sound farming practices.

But in *Wildness: Relations of People and Place*, Gavin Van Horn and John Hausdoerffer pull together a series of essays by ecologists, farmers, scientists, environmental advocates, and activists to examine the en-tangled relationship between our wilderness areas and the human-made world.[21] Can the world survive without the Boundary Waters Canoe Area or the Maroon Bells Wilderness? Perhaps a better question, many of the writers argue, is whether humans can survive without "wildness" and what it contributes toward creating a working ecosystem, whether it be in a remote mountain range, or in our parks and cities, as well as on our farms.

"Wildness is not an all-or-nothing proposition. There are variations, ranging from the sunflower pushing through a crack in a city alley to the cultivated soils of a watershed cooperative to thousands of acres of multi-generational forestlands," writes Van Horn.[22]

Designated wilderness areas are important, but in farm country we need to blur the lines between natural and agricultural, and between producer and consumer. Buying certified organic milk and grass-fed beef, or patronizing the local farmers' market helps, but it needs to go one step beyond to where people consider eating, as Wendell Berry puts it, "an agricultural act."[23] On one visit to the Loretta and Martin Jaus farm, I started to see a little bit of this kind of ethic taking shape.

A Bountiful Table

On an unpleasantly hot Saturday in mid-June—a one-two punch of humidity and haze had triggered an air quality warning in the region— a couple dozen people gathered in a loose group between a newly sprouted cornfield and a windbreak. This field day was taking place on the Jaus farm, which wasn't unusual—they often host people who are interested in learning more about their techniques for producing organic milk. What was out of the ordinary was what the field day participants carried. Almost to a person, they were armed with binoculars and field guides. Some were even equipped with smart phone apps that can help identify birds via their calls.

This tour was being sponsored by the Minnesota chapter of the Audubon Society as well as Birds and Beans, a marketer of shade-grown coffee, and Organic Valley, the organic dairy cooperative the Jauses sell their milk to. The nature of the sponsorship was no accident. Titled "Food, Farms, and Feathers: How Sustainable Agriculture Supports Healthy Ecosystems," this field day was set up to help birding enthusiasts and other environmentalists connect more birds on their life lists to what kind of coffee or milk they consume.

"Our decisions to consume products in an ecologically sustainable way makes a difference," the Fish and Wildlife Service's Tom Will told the crowd. "Our choices make a difference."

This was an excellent opportunity to make that connection. Will explained how songbirds such as bobolinks and dickcissels that make their summer home on the Jaus farm fly thousands of miles to spend their winters in South American countries like Columbia or Nicaragua. "And every place they visit contributes to their wellbeing and health," he explained. "So it's critical, for example, for those birds to find a gem of a place like this farm where they can respond to organic farming practices, healthy farming practices, lots of energy, lots of insects to feed their young. And that fuels their ability to make that long-distance journey all the way back to South America and hopefully find a similar environment there where they'll encounter a healthy environment relatively free of pesticides and things that would make it more difficult for them to return to us the next year."

Roughly half a century ago, coffee growers in the Southern Hemisphere started cutting down forests to provide coffee bushes with more full sun. This caused the bushes to produce beans faster and was more profitable for the farmers, but it also caused serious erosion and decimated the overstory wintering songbirds require. The farmers, as well as other people in these communities, also had to deal with the inevitable chemical regimen that comes with replacing a diverse habitat with a monoculture. In the 1990s, organizations and businesses began certifying coffee that was grown with the overstory intact. Called "shade-grown" coffee, this burgeoning market provides a price premium for farmers who are willing to grow their coffee without removing vast tracts of forest habitat.[24]

Birds & Beans is one of the companies importing this bird-friendly coffee. This is a good example of putting your money where your ethics are. Anyone who enjoys seeing Baltimore orioles in their backyard should be drinking shade-grown coffee. It's a clear and simple link between our actions here and results thousands of miles south. But what hasn't always been so clear is how our consumption of products produced right in our own backyard impacts these same birds. Since they summer here, such a connection would seem to make sense. But let's face it, midwestern farmland is not as charismatic as a South American forest.

Part of the Jauses' responsibility on this particular day was to fill in the tour participants on the midwestern end of the cycle, and to show that working farms like theirs are just as valuable in supporting healthy bird habitat as an ecologically mindful coffee operation in Nicaragua. After brief introductions, we took a hay wagon tour of the farm, during which Martin described their method of producing milk, interrupting himself periodically to identify a bird. As we went at a parade pace by the orderly pasture paddocks bordered by windbreaks, the farmer explained their grazing and chemical-free cropping systems. He also gave a little history lesson. In 1877 this farm was homesteaded by Martin's great-grandfather. In the 1960s, modern drainage ditches came in and replaced bird habitat, he said with an air of melancholy. It wouldn't be the last time he reminded the field tour participants of a wilder time with a sense of sadness. He pointed to the south where a lake used to be. "My dad used to go fishing there," he said. "That's all gone now. The landscape has changed."

But he made it clear we can't completely blame the lack of wildlife habitat on trends set in motion by decisions made decades ago. "What did you see on your way here in the road ditches?" Martin asked the group rhetorically. "Mowing. It's prime nesting season. That's something that really bothers me." Another result of the takeover of the midwestern rural landscape by corn and soybeans is that farmers desperate for livestock forage are baling up the grasses found in road ditches.

During the tour, Martin returned repeatedly to the theme of historical, and not so historical, landscape-level destruction. But the tour participants also saw what one committed farm family can do to bring their own piece of property back to life in a matter of decades. Such signs of life were everywhere we looked: dickcissels clinging to waving grasses like pole vaulters caught in mid-arch, making their dry, insect-like call; tree swallows swooping through the grazing paddocks, picking insects out of the air; a young red-tailed hawk riding thermals above the fields; a clay-colored sparrow with its *zhee, zhee, zhee* song; raucous red-winged blackbirds dominating the cattails; eastern kingbirds balancing on barbed wire; and male Baltimore orioles, flashing their orange feathers as they weaved in and out of trees.

We drove past beehives that honey producers have set up near the Jaus farmstead next to some grazing paddocks and a windbreak, a reminder that this farm is good for pollinators as well. At one point in the tour, Martin reminded us of another public service the farm provides. They have a "living snow fence" of green ash and red cedar. Since they put it in, the county road department has not had problems with the byway drifting in. Yes, this farm had been brought back to life. But there are limits. At one point as we rode along a gravel road past a drainage ditch, Martin talked about how they plugged tile lines to reclaim an eleven-acre marsh. But it can be hard to do that beyond a limited basis, since one farm's drainage infrastructure is sometimes intertwined with a neighbor's system, much like the storm sewers in an urban or suburban area.

"It's difficult to bring back a wetland," Martin said in his typical understated way.

While on the tour, the participants, which included a birding columnist for a city newspaper, had a host of questions:

- How long does it take to become certified organic? (There is a three-year transition period.)
- How do you control weeds without chemicals? (A flame weeder and mechanical cultivation.)
- How do you deal with being surrounded by genetically modified crops? (Plant later than the neighbors to avoid cross-pollination.)
- What did this area once look like? (When it was homesteaded, it was all marsh and tallgrass prairie.)

After the hay-ride tour, we gathered at the field road next to the cornfield and passed through the windbreak, where we walked a winding path through a dense stand of woody wildlife plantings. We emerged into an open area next to a long, rectangular pond. A white open-sided tent had been set up with tables and folding chairs. At one end was a locally sourced meal laid out buffet style. North of us were grazing paddocks. Orioles darted in and out among the trees near the pond. The well-kept barn and farmstead were to the west. The set-up made it look like the editors at *Birds and Blooms, Bon Appétit,* and *Successful Farming* magazines had decided to save money and hire one photo stylist for the day.

A chalkboard next to the buffet table listed the menu:

- Beef tenderloin with creamed mustard greens and arugula pesto.
- Heirloom bean salad with arugula, micro-buckwheat, and mustard vinaigrette.
- Beet, dill, cilantro farro tabbouleh.
- Summer greens with radish and turnip and honey vinaigrette.
- Cornbread muffins.

And, because it was that time of year in the Midwest, the meal was topped off with locally sourced rhubarb pie.

A local restaurant had prepared the food, and the owner talked to us about the importance of sourcing locally to "make connections" like we saw here. We were also encouraged to keep in mind the $3 million in salaries her restaurant pumps into the local economy. Aimee Haag, who runs Rebel Soil Farm with her husband Andy Temple, had provided the vegetables and she gave a short presentation on their operation. The young farmer got emotional talking about the importance of sourcing locally and how a local foods restaurant can contribute to the community's economy.

"When I see all you can do on these acres it makes me excited about what we can do on our four acres," Haag said to the Jauses.

After lunch, Kristin Hall, the conservation manager for Audubon Minnesota, talked about the "powerful changes" we can make with our food choices. "Not all of us can buy a farm," she said, re-emphasizing that what we can do is support this kind of farming with our buying habits. "Not every farm can be like this, but there's hope."

She then introduced a panel discussion involving Will and Loretta, as well as Katie Fallon, author of *Cerulean Blues: A Personal Search for a Vanishing Songbird* and the co-founder of the Avian Conservation Center of Appalachia. Will and Fallon, who have both studied birds in the Southern Hemisphere, reinforced the connections between a midwestern dairy farm, South American tropical forests, and the food choices we make. Will talked excitedly about new developments in ornithology and about cutting-edge information that has been gathered from tiny backpacks on migrating birds. As they spoke, the dickcissels in the nearby grazing paddocks sang incessantly—*dick, dick, dick . . . cissel, dick, dick, dick . . . cissel.*

Fallon talked about the human connections that migrating songbirds can forge. On cue, a Baltimore oriole started whistling clearly from near the pond.

"That Baltimore oriole could have been seen by somebody at a coffee farm in Nicaragua," said Fallon. "Some person picking coffee beans could have heard that fluty song and looked up and said, 'Wow, that's a really cool bird.' So just seeing that Baltimore oriole links you to someone thousands of miles away who could have looked at that bird and had the same reaction—'Wow! Look at that thing.'"

By the time Loretta got up to speak, I realized she had a tougher case to make. During the picnic, an Organic Valley tanker truck had driven into the nearby farmyard and completed its once-a-week pickup of the Jaus milk, draining the bulk tank in the eighty-eight-year-old stone barn. As cues go, Fallon had her oriole, Loretta had her truck. "I heard the rumble of the truck go by and I was thinking about how much simpler our life would be if we just went for the biggest milk check," she said. But by being certified organic, they don't hoe the easy row. And in fact, they have chosen an even tougher path than many of their organic colleagues. Besides the ban on toxic chemicals and petroleum-based fertilizer, among other things, organic certification has certain stipulations about retaining wildlife habitat and making sure cows get the bulk of their feed during the growing season from well-managed pastures. That's important. The chemical restrictions and pasture requirements alone make organic farms much friendlier to wildlife and pollinators, not to mention the soil universe.

But organic certification does not require a farmer to restore a wetland or put up bluebird boxes, as the Jauses have. In fact, surveys show that shoppers consistently choose organic products because they think consuming them is healthier for their family.[25] Whenever a food scare flares up in the industrialized system, organic sales spike. That's great, and it's supporting farmers like the Jauses. But let's face it, the consuming public is fickle. As soon as some sharp marketer convinces them that a different kind of food is healthier, they lurch to that side of the ship.

And then there's the fact that the Organic Valley truck will be driving through mile after mile of monocropped fields that come autumn will be stripped of all plant life as a result of economic and policy based choices. Granted, several participants in this small tour pledged to purchase organic dairy products from now on, but buying "green" doesn't have the same long-term, generational impact that adopting a new ethic toward the land does, an ethic that means you will support a certain kind of farming in as many ways as possible—from the decisions made in the grocery store to those made in the voting booth.

So what did Loretta do? She told a story. It was the one I shared in chapter 1 about the time the grasshoppers got inebriated feeding on the Jauses' small grains because of the high biological activity in the farmers' soils. It's a perfect story to tell to just about any crowd, since it contains all the great elements of a narrative: drama, tragedy, mystery, humor— even an aha moment. It also conveys an important message that even when we think we know everything, we really don't. Most important, those grasshoppers, those plants, and that soil were making it clear that the land is part of a whole. Farmers, and by extension the people they feed, are only fooling themselves if they think they are not members of that same community.

"We learned to watch for things like that," Loretta said at the conclusion of her story. "We have a lot of faith in the land. We were definitely making the management decisions on the farm that we did because we felt it was the right thing to do, without really understanding the full ramifications and potential of what we were doing. What the grasshopper story taught us was that we can run the farm, and that we should run the farm, with respect for the principles that are operating the natural systems and that in the end it will pay off in big ways, not only in terms of success of the farm, but in terms of the other living things we share the farm with."

She stopped and let the story soak in while the rest of us glanced around, perhaps seeing the land in a deeper way.

11

Wildly Optimistic

It's Hard to Be a Pessimist in a Land of New Possibilities

I've made it clear that this book is no attempt to provide some sort of how-to guide for wildly successful farming. Rather, this is more of a report from the frontlines of a phenomenon that I hope can serve as an inspiration—for other farmers, as well as policymakers, educators, scientists, conservationists, and anyone else who cares about the relationship between land, people, and food production. While I've shared some observations on what makes a wildly successful farm, I avoided providing a checklist of requirements for being ecologically correct.

However, there is one common trait shared by all the farmers featured here: optimism. That may come as a surprising characteristic for these farmers, given the debate over "technological pessimism" and "technological optimism." Basically, technological optimists believe that no matter what the problem—climate change, water pollution, energy production, or feeding people—there is or will be a technological fix for it. The technological pessimists tend to operate under the assumption that we simply cannot "tech" our way out of an irrefutable fact: the Earth has a finite amount of natural resources.[1]

Technological optimists often maintain that if you question even in the slightest the ability of machines, the digital world, chemicals, and genetically modified organisms to solve our problems, then you are outright opposed to *all* technology and human innovation. As it happens,

Martin Jaus (*second from left*) leads a birding tour of his dairy farm. "If those sounds weren't there, we would consider ourselves a failure," he says.

when it comes to farming, mainstream agriculture tends to consider sustainable, organic, and otherwise "alternative" farmers as technological pessimists to the core. Listen to this kind of talk long enough and unconventional farmers come off as a dour bunch of Luddites scratching at the soil with hoes they fashioned from windfall tree branches. The reality is that such farmers are fully embracing everything from high-tech fencing and computer software to cutting edge soil health tests and sophisticated plant breeding. Yes, many are even using herbicides and other agrichemicals. I recently spent time with a pair of beginning farmers striving to be "wildly successful" who are putting their academic training in calculus, plant pathology, and electronics engineering to use developing a better pasture system.

I see wildly successful farmers as "optimistic realists." Sure, they question the ability of technology to solve all our problems, but that doesn't make them cynical and negative about innovation. Far from it. They tend to put their trust in the hands of a different driver of change—in this case a working ecosystem.

"If those sounds weren't there, we would consider ourselves a failure," farmer Martin Jaus told me while standing next to a mix of wetland and prairie habitat he and his wife Loretta restored on prime crop ground. Taken at face value, that might appear to be a pessimistic statement.

But *because* the Jauses have put their trust in the presence of noisy wild-life as their gauge of success, they have taken steps to ensure they meet such a standard. That has resulted in spillover benefits all across the farm—for the land as well as the people residing on it.

Optimistic realists are mindful of the place technology should occupy in human endeavors, whether they involve farming, manufacturing, or education. When we become, to paraphrase Henry David Thoreau, tools of our tools, then we are vulnerable to the vicious cycle of regrettable substitution. When tools are put in their proper place, then wildly successful farmers can open themselves up to the multiple benefits of a more holistic, integrated approach. Indiana farmer Dan DeSutter was able to stop relying on a steel ripper to break up compacted soil because he knew the implement was just an isolated tool and didn't provide the numerous additional ecological/agronomic services that cover crops do.

Some may argue that by placing their trust in the ways of the wild, farmers are abdicating control over their own destiny in a way that's no better than allowing technology to call the shots. But this book describes farmers who are innovative, ambitious, and constantly on the lookout for a better way to do things—the opposite of being passive recipients of whatever life brings their way.

Consider the current excitement over soil health. From North Dakota to Indiana I've been on farms where, for the first time in decades, their owner-operators feel somewhat in control of their future. Why? When a farm is almost completely reliant on petroleum-based inputs, combined with technology developed in Monsanto's laboratories, events far from the land determine that farmer's destiny. Just putting in more grueling hours, which many do, isn't enough. War in the Middle East can disrupt the flow of oil; yet one more consolidation in the biotechnology sector can limit the availability of affordable seed. But building soil health starts and ends with a farm's local terra firma, literally from beneath the ground up. These farmers are striving for a situation that eventually makes their soil self-sufficient, not increasingly dependent on outside inputs and knowledge. That makes those farmers partners in the process. Granted, they may have a minority ownership in this partnership, but at least they aren't occupying the bystander role many conventional farmers are relegated to.

The interest in building soil health highlights another myth attached to farmers who question the primacy of technological fixes—that such doubters are inherently anti-science. Wildly successful farmers are natural observers and have an almost unquenchable thirst for knowledge—two characteristics of a good scientist. I find the increasing number of alliances developing between farmers, scientists, and environmental experts to be particularly exciting. The best of these partnerships go beyond the traditional paradigm of the land grant scientist passing knowledge from on high down to the farmer. Like North Dakota's Burleigh County Soil Health Team, the parties involved are learning from each other and share the realization that there are limits to what we can know. Such linkages forged by humbleness can make for a truly egalitarian relationship, the kind that fosters real, long-term change.

An optimistic outlook doesn't mean ignoring the realities of trying to make a living on the land. All farmers face daunting obstacles, no matter what management philosophy they are practicing. But the wildly successful farmers I've seen in action seem to be better able to roll with the punches with some positive vision for the future still intact. They know, for better or worse, that nature truly does have the final say.

One can't eat a pretty view, but it feeds us in other ways. After spending even a short time with these farmers, I often notice how their language has become saturated with references to the various workings of the landscape. They have allowed nature to guide their farming decisions, and they communicate their greatest joys by referring to how a healthy ecosystem makes its presence known. Martin Jaus actually doesn't savor describing in detail his grazing system or how he rotates crops. But as soon as the talk turns to his latest sighting of a hawk on a hayfield, he is ecologically animated. Jan Libbey's passion for birds popping up literally in the middle of a workshop on farm finances is a delight to behold. Such descriptions of nature don't always have to be of the charismatic variety that make for great photos like the ones Phil Specht takes. When Brooke Knisley talks excitedly about her soil coming to life in the middle of a monocultural desert, one can picture a world that rivals the Amazonian jungle in terms of biodiversity.

Such reflections are just as fun to listen to as they are for the farmers to relate. And that's another reason I know these people are optimists. To them, each day brings a fresh opportunity to gather material for a new story inspired by the land.

Notes

Introduction

1. Louisiana Universities Marine Consortium, "2017 Shelfwide Cruise: July 24–July 31," https://gulfhypoxia.net (accessed August 2, 2017).

2. Michael Egan, *Barry Commoner and the Science of Survival: The Remaking of American Environmentalism* (Cambridge: MIT Press, 2007), 126–27.

3. R. E. Turner and N. N. Rabalais, "Linking Landscape and Water Quality in the Mississippi River Basin for 200 Years," *BioScience* 53, no. 6 (2003): 563–72.

4. U.S. Environmental Protection Agency, "Mississippi River/Gulf of Mexico Hypoxia Task Force," https://www.epa.gov/ms-htf (accessed March 26, 2017).

5. Iowa Department of Natural Resources Geological Survey Bureau, *Big Spring Retrospective*, Fact Sheet, Ambient Monitoring Program, Iowa Department of Natural Resources (Iowa City: State of Iowa, 2002), 1–4.

6. E. McLellan, D. Robertson, K. Schilling, M. Tomer, J. Kostel, D. Smith, and K. King, "Reducing Nitrogen Export from the Corn Belt to the Gulf of Mexico: Agricultural Strategies for Remediating Hypoxia," *Journal of the American Water Resources Association* 51, no. 1 (2015): 263–89.

7. MacKenzie Elmer, "Des Moines Water Works Won't Appeal Lawsuit," *Des Moines Register*, April 11, 2017, 1.

8. Tom Feran, "Toxin in Lake Erie Puts Toledo Drinking Water on 'Watch,'" *The Plain Dealer*, July 30, 2015.

9. U.S. Environmental Protection Agency, "Nonpoint Source: Agriculture," https://www.epa.gov/nps/nonpoint-source-agriculture (accessed July 28, 2017).

10. E. Sinha, A. M. Michalak, and V. Balaji, "Eutrophication Will Increase during the 21st Century as a Result of Precipitation Changes," *Science* 357, no. 6349 (2017): 405–8.

11. Dan Undersander, Beth Albert, Dennis Cosgrove, Dennis Johnson, and Paul Peterson, *Pastures for Profit: A Guide to Rotational Grazing (A3529)*, How-to Guide, Extension Service, University of Wisconsin/University of Minnesota (Madison: Cooperative Extension Publishing, University of Wisconsin Extension Service, 2002), 1–38.

12. Jennifer Taylor and Steve Near, "How Does Managed Grazing Affect Wisconsin's Environment?," University of Wisconsin–Madison Center for Integrated Agricultural Systems, October 2008, http://www.cias.wisc.edu/how-does-managed-grazing-affect-wisconsins-environment/ (accessed March 28, 2017). W. R. Teague, S. Apfelbaum, R. Lal, U. P. Kreuter, J. Rowntree, C. A. Davies, and R. Conser, "The Role of Ruminants in Reducing Agriculture's Carbon Footprint in North America," *Journal of Soil and Water Conservation* 71, no. 2 (2016): 156–64.

13. Tom Kriegl and Ruth McNair, "Pastures of Plenty: Financial Performance of Wisconsin Grazing Dairy Farms," University of Wisconsin–Madison Center for Integrated Agricultural Systems, February 2005, http://www.cias.wisc.edu/pastures-of-plenty-financial-performance-of-wisconsin-grazing-dairy-farms/ (accessed March 28, 2007).

14. David R. Huggins and John P. Reganold, "No-Till: How Farmers Are Saving the Soil by Parking Their Plows," *Scientific American* (July 2008): 70–77.

15. Tom Philpott, "A Reflection on the Lasting Legacy of 1970s USDA Secretary Earl Butz," February 8, 2008, http://grist.org/article/the-butz-stops-here/ (accessed March 29, 2017).

16. Aldo Leopold, *A Sand County Almanac and Sketches Here and There* (New York: Oxford University Press, 1949), 224–25.

17. Kristin Ohlson, *The Soil Will Save Us: How Scientists, Farmers, and Foodies Are Healing the Soil to Save the Planet* (New York: Rodale, 2014), 78–112.

18. Leopold, *A Sand County Almanac*, 214.

19. Aldo Leopold, *For the Health of the Land: Previously Unpublished Essays and Other Writings*, edited by J. Baird Callicott and Eric T. Freyfogle (Washington, DC: Island Press, 1999), 161–175.

20. Dana L. Jackson and Laura L. Jackson, *The Farm as Natural Habitat: Reconnecting Food Systems with Ecosystems* (Washington, DC: Island Press, 2002), 11–26.

21. Minnesota Department of Natural Resources, *Minnesota Wetlands Conservation Plan, Version 1.02, 1997*, DNR Ecological Services Section (St. Paul: Minnesota Department of Natural Resources 1997), 108.

22. Leopold, *For the Health of the Land*, 166.

23. Laura Lengnick, *Resilient Agriculture: Cultivating Food Systems for a Changing Climate* (Gabriola Island, BC: New Society Publishers, 2015), 41–62.

24. United States Department of Agriculture National Agricultural Statistics Service, "Farms and Farmland: Numbers, Acreage, Ownership, and Use," in

2012 Census of Agriculture Highlights, September 2014, https://www.agcensus
.usda.gov/Publications/2012/Online_Resources/Highlights/Farms_and_
Farmland/Highlights_Farms_and_Farmland.pdf (accessed March 29, 2017).

25. Brian DeVore, "Why Do They Do It?," in *The Farm as Natural Habitat:
Reconnecting Food Systems with Ecosystems*, edited by Dana L. Jackson and Laura L.
Jackson, 107–18 (Washington, DC: Island Press, 2002).

26. Leopold, *For the Health of the Land*, 164.

Chapter 1. Beyond the Pond

1. J. Fischer, B. Brosi, G. C. Daily, P. R. Ehrlich, R. Goldman, J. Gold-
stein, and D. B. Lindenmayer, "Should Agricultural Policies Encourage Land
Sparing or Wildlife-Friendly Farming?," *Frontiers in Ecology and the Environment* 6,
no. 7 (2008): 380–85.

2. USDA Economic Research Service, "Most Farmers Receive Off-Farm
Income, but Small-Scale Operators Depend on It," in *Farming and Farm Income*,
November 30, 2016, https://www.ers.usda.gov/data-products/chart-gallery
/gallery/chart-detail/?chartId=58426 (accessed July 28, 2017).

3. Wild Farm Alliance, "Biodiversity Compliance Assessment in Organic
Agricultural Systems," January 2008, http://www.wildfarmalliance.org/bio
diversity (accessed July 16, 2017).

4. Aldo Leopold, *Round River: From the Journals of Aldo Leopold*, edited by
Luna B. Leopold (New York: Oxford University Press, 1993), 145–46.

5. Minnesota Pollution Control Agency, "Minnesota River Basin," https://
www.pca.state.mn.us/water/minnesota-river-basin (accessed April 13, 2017).

6. L. Knobeloch, B. Salna, A. Hogan, J. Postle, and H. Anderson, "Blue
Babies and Nitrate-Contaminated Well Water," *Environmental Health Perspectives*
108, no. 7 (2000): 675–78.

Chapter 2. A Place in the Country

1. Jean Cutler Prior, *A Regional Guide to Iowa Landforms*, Guidebook, Iowa
Geological Survey (Iowa City: State of Iowa, 1976), 41–44.

2. K. J. Donham, S. Wing, D. Osterberg, J. L. Flora, C. Hodne, K. M. Thu,
and P. S. Thorne, "Community Health and Socioeconomic Issues Surrounding
Concentrated Animal Feeding Operations," *Environmental Health Perspectives* 115,
no. 2 (2007): 317–20.

3. National Research Council of the National Academies, *Status of Pollinators
in North America*, Status Report (Washington, DC: National Academies Press,
2007), 312.

4. Brian DeVore, "A Sticky Situation for Pollinators," *Minnesota Conservation Volunteer* (July–August 2009): 8–17.

5. D. van Engelsdorp, J. D. Evans, C. Saegerman, C. Mullin, E. Haubruge,

B. K. Nguyen, and M. Frazier, "Colony Collapse Disorder: A Descriptive Study," *PLoS ONE* 4, no. 8 (2009): e6481.

6. U.S. Fish and Wildlife Service, "Fact Sheet: Rusty Patched Bumble Bee (Bombus affinis)," *U.S. Fish and Wildlife Service Endangered Species,* January 10, 2017, https://www.fws.gov/midwest/endangered/insects/rpbb/factsheetrpbb .html (accessed April 13, 2017).

7. M. C. Donaldson-Matasci and A. Dornhaus, "How Habitat Affects the Benefits of Communication in Collectively Foraging Honey Bees," *Behavioral Ecology and Sociobiology* 66, no. 4 (2012): 583–92.

8. Y. J. Cardoza, G. K. Harris, and C. M. Grozinger, "Effects of Soil Quality Enhancement on Pollinator-Plant Interactions," *Psyche: A Journal of Entomology* (2012): 1–8.

Chapter 3. Blurring the Boundaries

1. Doug Smith, "Losses of Habitat as Grasslands Plowed Mean Trouble for Hunters and Wildlife," *Minneapolis Star Tribune,* September 28, 2013, C8.

2. C. K. Wright and M. C. Wimberly, "Recent Land Use Change in the Western Corn Belt Threatens Grasslands and Wetlands," *Proceedings of the National Academy of Sciences* 110, no. 10 (2013): 4134–39.

3. C. K. Wright, B. Larson, T. J. Lark, and H. K. Gibbs, "Recent Grassland Losses Are Concentrated around U.S. Ethanol Refineries," *Environmental Research Letters* 12, no. 4 (2017): 1–16.

4. T. J. Lark, M. J. Gibbs, and H. K. Salmon, "Cropland Expansion Outpaces Agricultural and Biofuel Policies in the United States," *Environmental Research Letters* 10, no. 4 (2015): 1–11.

5. Ibid.

6. Kelly April Tyrrell, "Plowing Prairies for Grains: Biofuel Crops Replace Grasslands Nationwide," University of Wisconsin–Madison, April 2, 2015, http://news.wisc.edu/plowing-prairies-for-grains-biofuel-crops-replace-grass lands-nationwide/ (accessed August 2, 2017).

7. Brian DeVore, "Grazing as a Public Good," *Land Stewardship Letter* 32, no. 1 (2014): 24–25.

8. Justin Pepper, *Conservation Grazing: An Opportunity for Land Trusts and Conservation Organizations,* Report, National Audubon Society (Arlington: The Pasture Project, Wallace Center at Winrock International, 2016), 12.

9. S. Swanson, S. Wyman, and C. Evans, "Practical Grazing Management to Maintain or Restore Riparian Functions and Values on Rangelands," *Journal of Rangeland Applications* 2 (2015): 1–28.

10. Minnesota Department of Natural Resources, *Minnesota Prairie Conservation Plan: A Habitat Plan for Native Prairie, Grassland, and Wetlands in the Prairie Region of Western Minnesota* (St. Paul: Minnesota Department of Natural Resources, 2011), 55.

11. Ibid.

12. United States Department of Agriculture Natural Resources Conservation Service, "Soil Biology: Key Educational Messages," University of Illinois Extension, 2017, https://extension.illinois.edu/soil/sb_mesg/sb_mesg.htm (accessed April 13, 2017).

13. Brian DeVore, "Hitting the Conservation Target: When It Comes to Making the Ag Landscape Healthier, How Much Is Enough?," *Land Stewardship Letter* 33, no. 1 (2015): 27–28.

Chapter 4. Brotherhood of the Bobolink

1. D. D. Smith, "Iowa Prairie: Original Extent and Loss, Preservation and Recovery Attempts," *Journal of the Iowa Academy of Science* 105, no. 3 (1998): 94–108.

2. Cornell Lab of Ornithology, "Cerulean Warbler," 2015, https://www.allaboutbirds.org/guide/Cerulean_Warbler/lifehistory (accessed April 13, 2017).

3. J. D. McCracken, "Where the Bobolinks Roam: The Plight of North America's Grassland Birds," *Biodiversity* 6, no. 3 (2005): 20–29.

4. Ward Laboratories, "Haney Soil Health Test Information," https://www.wardlab.com/haney-info.php (accessed April 14, 2017).

5. N. Fierer, J. Ladau, J. C. Clemente, J. W. Leff, S. M. Owens, K. S. Pollard, R. Knight, J. A. Gilbert, and R. L. McCulley, "Reconstructing the Microbial Diversity and Function of Pre-Agricultural Tallgrass Prairie Soils in the United States," *Science* 342, no. 6158 (2013): 621–24.

6. Roger Tory Peterson, *A Field Guide to the Birds East of the Rockies* (Boston: Houghton Mifflin, 1980), 256–57.

7. Aldo Leopold, *A Sand County Almanac: And Sketches Here and There* (New York: Oxford University Press, 1949), 130–32.

Chapter 5. Raising Expectations

1. R. J. Manlay, C. Feller, and M. J. Swift, "Historical Evolution of Soil Organic Matter Concepts and Their Relationships with the Fertility and Sustainability of Cropping Systems," *Agriculture Ecosystems and Environment* 119, no. 3–4 (2007): 217–33.

2. F. R. Troeh and L. M. Thompson, *Soils and Soil Fertility*, 6th ed. (Ames, IA: Blackwell, 2005), 105–26.

3. Ray R. Weil and Nyle C. Brady, *The Nature and Properties of Soils*, 15th ed. (Washington, DC: Pearson, 2017), 1071.

4. Burleigh County Soil Conservation District, http://www.bcscd.com/store/pc/home.asp (accessed July 30, 2017).

5. D. Tilman, P. B. Reich, and J. M. H. Knops, "Biodiversity and Ecosystem Stability in a Decade-Long Grassland Experiment," *Nature* 441 (2006): 629–32.

6. J. J. Hoorman and R. Islam, "Understanding Soil Microbes and Nutrient Recycling," Ohio State University Extension, September 7, 2010, http://ohioline.osu.edu/factsheet/SAG-16 (accessed April 13, 2017).

7. Amber Dance, "Soil Ecology: What Lies Beneath," *Nature* 455 (October 8, 2008): 724–25.

8. U.S. Department of Agriculture Natural Resources Conservation Service, "Role of Soil Organic Matter," https://www.nrcs.usda.gov/wps/portal /nrcs/detailfull/soils/health/mgnt/?cid=nrcs142p2_053859 (accessed August 2, 2017).

9. U.S. Department of Agriculture Natural Resources Conservation Service, "Manage for Soil Carbon," https://www.nrcs.usda.gov/wps/portal /nrcs/detailfull/soils/health/mgnt/?cid=stelprdb1237584 (accessed August 2, 2017).

10. Allan Savory and Jody Butterfield, *Holistic Management: A Commonsense Revolution to Restore Our Environment*, 3rd ed. (Washington, DC: Island Press, 2016), 532.

11. Christopher Ketcham, "Allan Savory's Holistic Management Theory Falls Short on Science," *Sierra* (March/April 2017).

Chapter 6. Feeding Innovation's Roots

1. Amber Dance, "Soil Ecology: What Lies Beneath," *Nature* 455 (October 8, 2008): 724–25.

2. North Central Sustainable Agriculture Research and Education, *Cover Crop Survey: Annual Report 2015–2016* (College Park: United States Department of Agriculture, 2016), 41.

3. David R. Montgomery and Anne Bikle, *The Hidden Half of Nature: The Microbial Roots of Life and Health* (New York: W. W. Norton, 2016), 71–72.

4. Vaclav Smil, *Enriching the Earth: Fritz Haber, Carl Bosch, and the Transformation of World Food Production* (Cambridge: MIT Press, 2001), 5–12.

5. Brian DeVore, "Nature's Backlash," in *The Farm as Natural Habitat: Reconnecting Food Systems with Ecosystems*, edited by Dana L. Jackson and Laura L. Jackson, 33–34 (Washington, DC: Island Press, 2002).

6. B. Ryan and N. Gross, *Acceptance and Diffusion of Hybrid Corn Seed in Two Iowa Communities*, Research Bulletin 372, Sociology Subsection: Economics and Sociology Section (Ames: Iowa State College of Agriculture and Mechanic Arts, 1950), 663–708.

7. Conservation Cropping Systems Initiative, "A 321% Return on Investment: Ken, Roy, and Rodney Rulon—Arcadia, Indiana," Fact Sheet, Economics of Conservation Series, Natural Resources Conservation Service (United States Department of Agriculture, 2016), 4.

8. G. J. Arbuckle, *Iowa Farm and Rural Life Poll: 2015 Summary Report*, Extension and Outreach, Iowa State University (Ames: Extension Community and Economic Development Publications, 2016), 3–10.

9. Tom Feran, "Toxin in Lake Erie Puts Toledo Drinking Water on 'Watch,'" *The Plain Dealer*, July 30, 2015.

10. Maryland Department of Agriculture. "Maryland's Cover Crop Program: Providing Grants to Help Farmers Protect the Chesapeake Bay," brochure, Office of Resource Conservation, http://mda.maryland.gov/resource_conservation/counties/Cover%20Crop%20Program%20Overview.pdf, 2.

11. John P. Reganold and David R. Huggins, "No-Till: How Farmers Are Saving the Soil by Parking Their Plows," *Scientific American Magazine*, June 30, 2008.

12. Aldo Leopold, *A Sand County Almanac: And Sketches Here and There* (New York: Oxford University Press, 1949), 194.

13. Ray R. Weil and Nyle C. Brady, *The Nature and Properties of Soils*, 15th ed. (Washington, DC: Pearson, 2017), 1071.

14. Montgomery and Bikle, *The Hidden Half of Nature*, 105–6.

Chapter 7. Wrapping Around the Wrinkles

1. Curt Meine, "The Edge of Anomaly," in *Wildness: Relations of People and Place*, edited by G. Van Horn and J. Hausdoerffer, 33–42 (Chicago: University of Chicago Press, 2017), 33–42.

2. Stanley W. Trimble, *Historical Agriculture and Soil Erosion in the Upper Mississippi Valley Hill Country* (Boca Raton, FL: CRC Press, 2012), 290.

3. Brian DeVore, "Eyes on the Perennial Prize," *Land Stewardship Letter* 22, no. 3 (2004): 14–15.

4. Arthur S. Hawkins, "Return to Coon Valley," in *The Farm as Natural Habitat: Reconnecting Food Systems with Ecosystems*, edited by Dana L. Jackson and Laura L. Jackson, 57–70 (Washington, DC: Island Press, 2002).

5. Defenders of Wildlife, *America's 10 Most Endangered National Wildlife Refuges 2004* (Washington, DC: Defenders of Wildlife, 2004), 15–16.

6. N. G. Perlut, A. M. Strong, T. M. Donovan, and N. J. Buckley, "Regional Population Viability of Grassland Songbirds: Effects of Agricultural Management," *Biological Conservation* 141, no. 12 (2008): 3139–51.

7. Brian DeVore, "When Farmers Shut Off the Machinery," in *The Farm as Natural Habitat: Reconnecting Food Systems with Ecosystems*, edited by Dana L. Jackson and Laura L. Jackson, 83–95 (Washington, DC: Island Press, 2002).

8. Aldo Leopold, *For the Health of the Land: Previously Unpublished Essays and Other Writings*, edited by J. Baird Callicott and Eric T. Freyfogle (Washington, DC: Island Press, 1999), 168–69.

9. D. Undersander, S. Temple, J. Bartlet, D. Sample, and L. Paine, *Grassland Birds: Fostering Habitats Using Rotational Grazing*, University of Wisconsin Extension Bulletin (Madison: University of Wisconsin Board of Regents, 2000), 9.

10. Cornell Lab of Ornithology, "Red-Headed Woodpecker," 2015, https://www.allaboutbirds.org/guide/Red-headed_Woodpecker/lifehistory (accessed April 13, 2017).

11. Minnesota Pollution Control Agency, "Report on Nitrogen in Surface

Water," June 26, 2013, https://www.pca.state.mn.us/featured/report-nitro
gen-surface-water (accessed April 13, 2017).

12. R. N. Lubowski, R. Claassen, and M. J. Roberts, "Agricultural Policy
Affects Land Use and the Environment," *Amber Waves*, September 2006, 28–33.

Chapter 8. Resiliency vs. Regret

1. Minnesota Pollution Control Agency, "Few Waters in Far SW Minnesota Meet Swimmable Fishable Standards," December 18, 2014, https://www
.pca.state.mn.us/featured/few-waters-far-sw-minnesota-meet-swimmable-fish
able-standards (accessed April 14, 2017).

2. N. Fierer, J. Ladau, J. C. Clemente, J. W. Leff, S. M. Owens, K. S.
Pollard, R. Knight, J. A. Gilbert, and R. L. McCulley, "Reconstructing the
Microbial Diversity and Function of Pre-Agricultural Tallgrass Prairie Soils in
the United States," *Science* 342, no. 6158 (2013): 621–24.

3. Mary Shelley, *Frankenstein, or The Modern Prometheus* [1818] (New York:
Signet Classic, 1963), 39–40.

4. Ian Urbina, "As OSHA Emphasizes Safety, Long-Term Health Risks
Fester," *New York Times*, March 30, 2013, 1A.

5. United States Department of Agriculture, "2012 U.S. Census of Agriculture," May 2, 2014, https://www.agcensus.usda.gov/Publications/2012/ (accessed April 14, 2017).

6. John P. Reganold and David R. Huggins, "No-Till: How Farmers Are
Saving the Soil by Parking Their Plows," *Scientific American Magazine*, June 30,
2008.

7. Claire O'Connor, *Soil Matters: How the Federal Crop Insurance Program
Should Be Reformed to Encourage Low-Risk Farming Methods with High-Reward Environmental Outcomes* (New York: Natural Resources Defense Council, 2013), 11–12.

8. United States Department of Agriculture Economic Research Service,
"Recent Trends in GE Adoption," November 3, 2016, https://www.ers.usda
.gov/data-products/adoption-of-genetically-engineered-crops-in-the-us/re
cent-trends-in-ge-adoption.aspx (accessed April 14, 2017).

9. William Neuman and Andrew Pollack, "Farmers Cope with Roundup-
Resistant Weeds," *New York Times*, May 3, 2010, 1A.

10. B. E. Tabashnik, T. Brévault, and Y. Carrière, "Insect Resistance to Bt
Crops: Lessons from the First Billion Acres," *Nature Biotechnology* 31 (2013): 510–21.

11. D. J. Sullivan, A. V. Vecchia, D. L. Lorenz, R. J. Gilliom, and J. D.
Martin, "Trends in Pesticide Concentrations in Corn-Belt Streams, 1996–2006,"
Scientific Investigations Report 2009–5132, U.S. Geological Survey, U.S. De-
partment of the Interior (2009), 75.

12. Herbicide Resistance Action Committee, "Guideline to the Management
of Herbicide Resistance," http://hracglobal.com/files/Management-of-Herbi
cide-Resistance.pdf (accessed April 14, 2017).

13. E. D. Perry, F. Ciliberto, D. A. Hennessy, and G. Moschini, "Genetically Engineered Crops and Pesticide Use in U.S. Maize and Soybeans," *Science Advances* 2, no. 8 (2016): 1–8.

14. J. P. Myers, M. N. Antoniou, B. Blumberg, L. Carroll, T. Colborn, L. G. Everett, and M. Hansen, "Concerns over Use of Glyphosate-Based Herbicides and Risks Associated with Exposures: A Consensus Statement," *Environmental Health* 15, no. 19 (2016): 1–30.

15. Land Stewardship Project, *Soil Health, Water and Climate Change: A Pocket Guide to What You Need to Know* (Minneapolis: Land Stewardship Project, 2017), 7–32.

16. S. Brantly, "Forest Grazing, Silvopasture, and Turning Livestock into the Woods," *Agroforestry Notes*, August, 2014, 1–4.

17. Brenda B. Lin, "Resilience in Agriculture through Crop Diversification: Adaptive Management for Environmental Change," *BioScience* 61, no. 3 (2011): 183–93.

18. K. L. Painter, "General Mills, Cascadian Farm Back Development of Kernza Wheatgrass," *Minneapolis Star Tribune*, March 8, 2017, B1.

19. A. S. Davis, J. D. Hill, C. A. Chase, A. M. Johanns, and M. Liebman, "Increasing Cropping System Diversity Balances Productivity, Profitability and Environmental Health," *PLoS ONE* 10 (2012): e47149.

20. Vaclav Smil, *Enriching the Earth: Fritz Haber, Carl Bosch, and the Transformation of World Food Production* (Cambridge: MIT Press, 2001), 5–12.

21. L. L. Jackson, D. R. Keeney, and E. M. Gilbert, "Swine Manure Management Plans in North-Central Iowa: Nutrient Loading and Policy Implications," *Journal of Soil and Water Conservation* 55, no. 2 (2000): 205–12.

22. J. Burkholder, B. Libra, P. Weyer, S. Heathcote, D. Kolpin, P. S. Thorne, and M. Wichman, "Impacts of Waste from Concentrated Animal Feeding Operations on Water Quality," *Environmental Health Perspectives* 115, no. 2 (2007): 308–12.

23. Davis et al., "Increasing Cropping System Diversity."

24. North Central Sustainable Agriculture Research and Education, *Cover Crop Survey: Annual Report 2015–2016* (College Park: United States Department of Agriculture, 2016), 41.

25. Wild Farm Alliance, "Biodiversity Resources," http://www.wildfarm alliance.org/biodiversity (accessed April 14, 2017).

26. J. J. Mainea and J. G. Boylesa, "Bats Initiate Vital Agroecological Interactions in Corn," *Proceedings of the National Academy of Sciences of the United States* 112, no. 40 (2015): 12438–43.

27. L. A. Morandin, R. F. Long, and C. Kremen, "Hedgerows Enhance Beneficial Insects on Adjacent Tomato Fields in an Intensive Agricultural Landscape," *Agriculture, Ecosystems and Environment* 189 (2014): 164–70.

28. D. S. Karp, S. Gennet, C. Kilonzoc, M. Partyka, N. Chaumont, E. R. Atwill, and C. Kremen, "Comanaging Fresh Produce for Nature Conservation

and Food Safety," *Proceedings of the National Academy of Sciences of the United States* 112, no. 35 (2015): 11126–31.

29. Iowa State University, "Prairie Strips: Small Changes, Big Impacts," Fact Sheet, June 2017, https://store.extension.iastate.edu/product/15221 (accessed July 17, 2017).

30. D. D. Smith, "Iowa Prairie: Original Extent and Loss, Preservation and Recovery Attempts," *Journal of the Iowa Academy of Science* 105, no. 3 (1998): 94–108.

31. D. E. Popper and F. Popper, "The Buffalo Commons: Its Antecedents and Their Implications," *Online Journal of Rural Research and Policy* 1, no. 6 (2006): 1–26.

32. E. McLellan, D. Robertson, K. Schilling, M. Tomer, J. Kostel, D. Smith, and K. King, "Reducing Nitrogen Export from the Corn Belt to the Gulf of Mexico: Agricultural Strategies for Remediating Hypoxia," *Journal of the American Water Resources Association* 51, no. 1 (2015): 263–89.

33. Iowa State University, "Science-Based Trials of Rowcrops Integrated with Prairie Strips," https://www.nrem.iastate.edu/research/STRIPS (accessed July 17, 2017).

34. Jennifer Hopwood, "Pollinators and Roadsides: Managing Roadsides for Bees and Butterflies," Fact Sheet (Portland: Xerces Society for Invertebrate Conservation, 2010), 8.

Chapter 9. Which Came First, the Farmer or the Ecologist?

1. V. A. Nuzzo, "Extent and Status of Midwest Oak Savanna: Presettlement and 1985," *Natural Areas Journal* 6 (1986): 6–36.

2. M. K. Leach and T. J. Givnish, "Identifying Highly Restorable Savanna Remnants," *Transactions of the Wisconsin Academy of Sciences, Arts and Letters* 86 (1998): 119–28.

3. Craig Childs, *Apocalyptic Planet: Field Guide to the Future of the Earth* (New York: Vintage Books, 2012), 186–87.

4. EcoSun Prairie Farms, http://www.ecosunprairiefarms.org (accessed December 20, 2017).

5. J. B. Ignatiuk and D. C. Duncan, "Nest Success of Ducks on Rotational and Season-Long Grazing Systems in Saskatchewan," *Wildlife Society Bulletin* 29, no. 1 (2001): 211–17.

Chapter 10. Hubs of Hope

1. University of Wisconsin Center for Cooperatives, "Cooperatives in the U.S.," http://www.uwcc.wisc.edu/whatisacoop/History (accessed July 25, 2017).

2. John Ikerd, "New Farm Bill and U.S. Trade Policy: Implications for Family Farms and Rural Communities," presented at "Grain Place" Farm

Tour and Seminar, Aurora, Nebraska, July 27, 2002, http://faculty.missouri
.edu/ikerdj/papers/FarmBill.pdf (accessed January 24, 2017).

3. S. Leonard, "Nature as Ally: An Interview with Wendell Berry," *Dissent Magazine*, Spring 2012.

4. Michael Pollan, *The Omnivore's Dilemma: A Natural History of Four Meals* (New York: Penguin Press, 2006), 34.

5. United States Department of Agriculture, "Farm Demographics—U.S. Farmers by Gender, Age, Race, Ethnicity, and More," *2012 USDA Census Highlights*, May 2014 https://www.agcensus.usda.gov/Publications/2012/Online_Resources/Highlights/Farm_Demographics/#average_age (accessed July 25, 2017).

6. W. R. Teague, S. Apfelbaum, R. Lal, U. P. Kreuter, J. Rowntree, C. A. Davies, and R. Conser, "The Role of Ruminants in Reducing Agriculture's Carbon Footprint in North America," *Journal of Soil and Water Conservation* 71, no. 2 (2016): 156–64.

7. C. Hribar, *Understanding Concentrated Animal Feeding Operations and Their Impact on Communities* (Bowling Green: National Association of Local Boards of Health, 2010), 7–8.

8. S. J. Vermeulen, B. M. Campbell, and J. S. I. Ingram, "Climate Change and Food Systems," *Annual Review of Environment and Resources* (2012): 195–222.

9. I. Shcherbaka, N. Millara, and G. P. Robertson, "Global Metaanalysis of the Nonlinear Response of Soil Nitrous Oxide (N2O) Emissions to Fertilizer Nitrogen," *Proceedings of the National Academy of Sciences of the United States of America* 111, no. 25 (2014): 9199–204.

10. Courtney White, *Grass, Soil, Hope: A Journey through Carbon Country* (White River Junction, VT: Chelsea Green Publishing, 2014), 272.

11. Teague et al., "The Role of Ruminants," 162.

12. R. Johnson and J. Monke, *What Is the Farm Bill?*, White Paper, Congressional Research Service (Washington, DC: U.S. Congress, 2017), 13.

13. C. K. Wright, B. Larson, T. J. Lark, and H. K. Gibbs, "Recent Grassland Losses Are Concentrated around U.S. Ethanol Refineries," *Environmental Research Letters* 12, no. 4 (2017): 1–16.

14. J. Marcotty and T. Kennedy, "In Minnesota's Farm Country, Clean Water Is Costly," *Minneapolis Star Tribune*, December 6, 2015, 1A.

15. M. J. Roberts, N. Key, and E. O'Donoghue, "Estimating the Extent of Moral Hazard in Crop Insurance Using Administrative Data," *Review of Agricultural Economics* 28, no. 3 (2006): 381–90.

16. "Climate Change: Better Management of Exposure to Potential Future Losses Is Needed for Federal Flood and Crop Insurance," GAO-15-28 (Washington, DC: U.S. Government Accounting Office, 2014), 48.

17. R. N. Lubowski, R. Claassen, and M. J. Roberts, "Agricultural Policy Affects Land Use and the Environment," *Amber Waves*, September 2006, 28–33.

18. "Stanley Milgram Quotes," http://www.goodreads.com/quotes /231190-ordinary-people-simply-doing-their-jobs-and-without-any-particular (accessed April 14, 2017).

19. Jedediah Purdy, *After Nature: A Politics for the Anthropocene* (Cambridge, MA: Harvard University Press, 2015), 228–55.

20. Ibid., 232–33.

21. Gavin Van Horn and John Hausdoerffer, eds., *Wildness: Relations of People and Place* (Chicago: University of Chicago Press, 2017).

22. Gavin Van Horn, "Into the Wildness," in *Wildness: Relations of People and Place*, edited by Gavin Van Horn and John Hausdoerffer, 4–5 (Chicago: University of Chicago Press, 2017).

23. Wendell Berry, *What Are People For?* (San Francisco: North Point Press, 1990), 145–46.

24. R. Rice and M. Bedoya, "The Ecological Benefits of Shade-Grown Coffee," *Smithsonian Migratory Bird Center*, Smithsonian Institution, September 2010, https://nationalzoo.si.edu/scbi/migratorybirds/coffee/bird_friendly /ecological-benefits-of-shade-grown-coffee.cfm (accessed July 25, 2017).

25. Organic Trade Association, "Organic Market Analysis," 2016, https:// www.ota.com/resources/market-analysis (accessed April 14, 2017).

Chapter 11. Wildly Optimistic

1. J. Hochschild, A. Crabill, and M. Sen, *Technology Optimism or Pessimism: How Trust in Science Shapes Policy Attitudes toward Genomic Science*, White Paper (Washington, DC: Brookings Institution, 2012), 1–16.

Index